Early Essays on Acupuncture and Moxa

Volume I. Acupuncture

Edited and published by Eric Serejski and John Howard

Frederick, MD 21702

http://iandi.us

ISBN 978-1-944175-68-9

Introduction to the Essays

The series on early essays on acupuncture and moxa consists of three volumes grouping the first English texts covering this topic. These essays are fundamental within the context of history and historiography of the field, clinical applications, and early explorations of the mechanisms involved. This series truly belongs to the shelves of practitioners and libraries of Oriental Medicine schools.

Volume 1

Dunglison, Robley. *Acupuncture* (1839).

Elliotson, John. *Acupuncture* (1832).

Morand, J. *Memoir on Acupuncturation* (1825). Translated by Franklin Bache.

Churchill, James Mors. *A Treatise on Acupuncturation* (1821).

Kaempfer, Engelbeit. *Of the Cure of the Colick by the Acupunctura; Moxa* (1727).

Volume 2

Larrey, D. J. *On the Use of Moxa as a Therapeutical Agent* (1822). Translated by Robley Dunglison.

Temple, William. *An Essay upon the Cure of the Gout by Moxa* (1677).

Early Essays on Acupuncture and Moxa

Volume 3

Boyle, James. *A Treatise on Moxa* (1825).

Wallace, William. *A Physiological Enquiry Respecting the Action of Moxa and Its Utility in Inveterate Cases* (1827).

Note: All editorial additions are placed within brackets.

Contents

Contents

1. Of the cure of the Colick by the Acupunctura; Moxa, Kaempfer, 1727

Of the cure of the Colick by the Acupunctura

The History of Japan: Together with a Description of the
Kingdom of Siam. 1727, 3:263-272

Engelbeit Kaempfer

Of the cure of the colick by the acupunctura or needle-pricking, as it is used by the Japanese.

Description of this distemper

The particular sort of Colick, which the Japanese call Senki,
is an endemial distemper of this populous Empire, and
withal so common, that there is scarce one in ten grown
persons, who hath not some time or other felt its attacks.
Thus far do the air, which is otherwise very healthful, the
climate, the way of life of the natives, their victuals and
drink jointly influence the human body, and dispose it to an
invasion of this distemper. Foreigners are no less subject to
it, than the natives, when once they are come to taste the
liquors of the Country. This we found to be too true by our
own sad experience, when upon our arrival in the Country
we endeavoured, as is usual amongst sea-faring people, to
wash away the memory of the dangers, we had been
exposed to in our tedious and difficult passage, by a

plentiful use of the cold beer of this Country, call'd Sakki. This beer is brewed out of rice to the strength and consistence of Spanish wines. It is of such a nature, that it should not be drank cold, but moderately warm, and out of dishes, after the manner of the natives. The name of Senki is not given indifferently to all Belly-achs, but only to that particular sort, which besides a most acute pain in the guts, occasions at the same time convulsions in the groins. For such is the nature and violence of this distemper, that all the membranes and muscles of the abdomen are convulsed by it. As to the cause of it, and of colicks in general, the natives are of opinion, that it is not at all a morbific matter lodged in the cavity of the guts, which, they say, would occasion but a very slight pain, but that the seat of it is in the membranous substance of some other part of the abdomen, as for instance of the muscles, the peritonaeum, the omentum, the mesentery, or the guts, and that by stagnating there it turns into a vapour, or rather into a very sharp sower spirit, as they express themselves, which distends, cuts and corrodes the membranes wherein it is lodged. Upon the same theory is grounded their method of cure: whenever this spirit is let out of the narrow prison it hath been confined to, and set at liberty, that very moment, they say, the pain which it hath occasioned by distending those sensible parts wherein it lay, must cease. Before I proceed farther, I must beg leave to observe, that instead of the Latin name Colica, which is sometimes wrongly given to this distemper,

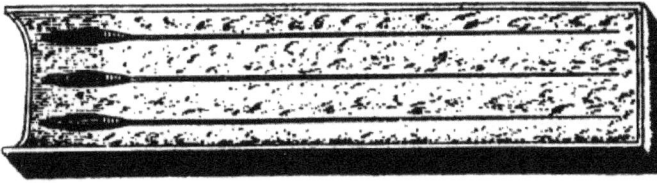

Figure 1. The silver needles in the case.

the gut, whence this name is derived, being frequently not so much as affected by it, the Brahmines chose rather to call it in their language, according to the opinion of the Chinese and Japanese, convulsions or spasms of the belly and guts. Some very particular symptoms of this endemial distemper of Japan are, that mimicking the hysteric affection, it often puts the patient under an apprehension of being suffocated, the whole region from the groins up to the false ribs, and higher, being strongly convulsed, that after it hath for a long time miserably tormented the patient, it will end in tumours, and swellings arising in several parts of the body, and attended with dangerous consequences, that particularly in men it will occasion a swelling in either of the testicles, which often suppurates and turns to an abscess, in women tubercula, or pustules in the anus and on the pudenda, commonly attended with the falling of the hair. It must be observed however, that both these tumors of the testicles, (which the Japanese call Sobi, and the patient afflicted with them Sobimotz) and the said pustules in the privities are likewise endemial distempers of this Country, and affect many, that have never lain sick of the colick.

Description of the needles

Before I proceed to shew, by what particular method the Japanese proceed in the cure of this distemper, which is by the needle, it may not be amiss to take notice, that there are two principal remedies in surgery, supposed to be equally successful in the cure and prevention of diseases, and which on this account are called in to assistance in

Figure 2. The covering of the said case.

these parts of the world by the healthful, as well as the sick, by regular Physicians and Quacks, by rich and poor. The Coraeans, Chinese and Japanese, all great admirers of antiquity, and scrupulous to excess in keeping up the ancient customs delivered down to them from their ancestors, unanimously pretend, that they were known in remotest ages, long before the invention of physick. Their very names indeed will appear terrible and shocking to the reader, they being no less, than fire and metal. And yet it must be owned in justice to the Japanese, that they are far from admitting of all that cruel, and, one may say, barbarous apparatus of our European surgery. Red hot irons, and that variety of cutting knives and other instruments requisite for our operations, a sight so terrible

to behold to the patient, and so shocking even to the
assistants, if they be not altogether destitute of all sense of
humanity and mercy, are things, which the Japanese are
entirely ignorant of. Their fire is but moderate, it hath
nothing to terrify the patient, it is such, as the very Gods of
the Country are not displeased to have burnt before them,
and in a word nothing else but a gently glowing tent of the
Plant, which bears the name of that celebrated Queen
Artemisia. So likewise the metals they make use of in their
operations of surgery, are the very noblest of all, the
ornament of royal palaces, the produce of sun and moon,
and, as the Philosophers pretend, richly imbued with the
qualities and virtues of those two celestial bodies. The
reader easily apprehends, that I mean, gold and silver, of
which they have needles made in a particular manner,
which are finely polished, and exceedingly proper to
perform the puncture in human bodies, and which are on
this account held in such an esteem by the natives, that
they constantly carry them along with them wherever they
go, as they do whole boxes of such other of their
instruments or curiosities, which they have a particular
value for, or are the most likely to want. The use and
application of both these remedies is a thing of such
consequence, that the very knowledge of the parts, which
are the most proper either to be burnt with the Moxa, or to
be pricked with the needles, is the object of a peculiar art,
the masters of which are called Tensasi, which is as much as
to say, touchers or searchers of the parts, because the main
business lies in the choice of the part, on which either of
these operations is to be performed. Those who manage
the needle, either pursuant to their own notions, or in
compliance with the patients desire, have the particular
name of Faritatte given them, which signifies Needle
Prickers. I now make haste to give a description of these

needles. It would be scarce possible to thrust a very thick needle into the body without some dangerous consequence or other: For this reason, the needles, whereby this operation is to be performed, must be exceeding small, made of either of the two metals abovementioned, so pure and fine as it is possible to get them, entirely separate from copper, and ductile. It is a particular art to temper these needles, and to bring them to a certain degree of hardness, requisite to make them fit for this operation, which art, although it be known but to very few persons, yet even those, who know it, are not allowed to make them without a particular license granted under the Imperial seal. There are two differing sorts of these needles, with regard to their structure. The first sort is made indifferently either of gold or silver; these are not unlike (as to their shape) to the bodkins, which our young boys at school spell withal, or the stylus's with which the Indians write, only they are smaller, about four inches long, thin, ending in a very sharp point, with a twisted

Figure 3. One of the gold needles taken out.

handle, in order to its being turn'd round or twisted with more ease. Instead of a box, they are kept in a small hammer, which is fitted up so, that on each side of the handle one of these needles may be conveniently lodged. This hammer is made of wild bulls-horns, finely polished, and is somewhat longer than the needle, with a compressed roundish head, wherein lies a piece of lead, to make it heavy. On that side, which touches the needle, in beating it into the body, it is defended by a piece of leather, commonly of a violet colour, and this to prevent, that in beating it should not leap up. The needles of the second sort are made only or silver, and are not unlike the first, as to their shape and length, but exceedingly small, with a short thick handle, which is striped or furrowed lengthways. They are kept several together in an oblong, square, wooden box, varnished without, with the bottom within covered with a piece of doth, in the woolly part of which the needles are stuck. For the satisfaction of those, who are curious in names, I have thought fit to take notice, that these two sorts of needles, and in general all needles, that are made use of in surgery, are called Uutsbarri, that is, turning or twisting needles. The needles of the second sort have the particular name of Fineribarri, which signifies the very same thing; and if the operation be performed, as is done frequently, through a small brass pipe, they are then called Fudabarri, that is channel'd needles. This pipe is about one third of an inch shorter than the needle, as big as a goose-quill, and serves to guide the needle, in order to make the puncture on any part of the human body so much the surer. These needles, with their cases, the hammer, and small pipe, are represented in Figure 1-Figure 5 wherein Figure 1 is the lower part of the case for the silver needles, with the needles lying in it. Figure 2. The covering of the said case. Figure 5. The brass pipe, which is to guide the operator in pricking. Figure 4. The hammer, with one of the gold-needles standing out a little way, and Figure 3 a gold-needle taken out.

But to come now to the operation itself, the same is performed after the following manner. The surgeon takes the needle near its point in his left hand, between the tip of the middle finger, and the nail of the forefinger, supported by the thumb, and so holds it toward the part which is to be pricked, and which must be first carefully examined, whether it be not perhaps a nerve, then with the hammer in his right hand, he gives it a knock, or two, just to thrust it through the hardish resistent outward skin. This done, he lays the hammer aside, and taking the handle of the needle between the extremities of the fore-finger and thumb, he twists it till the point runs into the body to that depth, which the rules of art require, being commonly half an inch, sometimes, but seldom, an inch or upwards, in short, till it runs into the place, where the cause of the pain and distemper is supposed to be hid, where he holds it, till the patient hath breathed once or twice, and then drawing it out, compresses the part with the finger, by this means, as it were, to squeeze out the vapour and spirit. The needles of the second sort are not knocked, but only twisted in, the operator holding them between the extremities of the thumb and middle finger: Those who are very dextrous at it, give it a knock with the fore-finger, laid upon the middle finger just to thrust it through the skin, and then they compleat the business by twisting; others make

Figure 4. The hammer with one of the gold needles standing out a little way.

use for this purpose of a pipe, such as above described, which is somewhat shorter than the needle, and will by this means stop it from running in too deep. The precepts and rules of this pricking art are very different, with regard chiefly to the hidden vapours, as the supposed cause of the distemper. Hence, when the operation is to be performed, a careful and circumspect

Figure 5. A brass pipe to guide the needles in pricking.

Physician must determine with all his attention and judgment, where and how deep they lie. The acupunctura is esteemed a very good remedy for those distempers, which are cured by burning with the Moxa, and the needle is to be applied nearly on the same places, and with the same cautions, as that Caustick; but of this more in my account of it. Even the common people will venture to apply the needle, meerly upon their own experience, and without the advice of an expert Tensasi, taking care only not to prick any nerves, tendons or considerable bloodvessels. Having premised thus much concerning the Acupunctura in general, it now remains to add a few words relating to its use in the cure of the colick in particular.

In order to cure the colick the Japanese perform this operation in the belly, in the region of the liver, making nine holes in three rows, disposed after the manner of a Parallelogram, at about half an inches distance from each other in grown persons, (Figure 6.) Each of these rows hath its peculiar name, as they are also made according to different rules. The first row is called Sioquan, and is made just beneath the ribs; the second row is called Tsiuquan, and claims the middle place between the navel and the cartilago mucronata, or ensiformal cartilage; the third is called Gecquan, and is made about half an inch above the

11

navel. I have been myself several times an eye-witness, that upon these three rows of holes, made according to the rules of art, and to a reasonable depth, the colick: Senki pains, as they call them, ceased almost in an instant, as if they had been charmed away.

Some endeavours have been made to cure this colick, by burning the patient with the Moxa, but upon trial this method hath not been found altogether so successful, as that of the Acupunctura. However it may not be amiss to take notice, that in this case the caustick must be applied to the belly, on both sides of the navel, about two inches from it. Both these places are called Tensu; they are famous for having numbers of causticks applied to them, and are known even to those, who do not practise this art. But of this more in another place.

Acupunctura Japonum

Figure 6. The Acupunctara, or needle-prickings of the Japanese, for curing the cholick.

To compleat this account, I must not forget to mention another remedy of pretended great efficacy, and frequently used by the common people in the colick, of which hitherto, as also in the cholera morbus, which is a very frequent and fatal distemper in this Country, in that belly-ach, which they call Saku, and which is likewise an endemial distemper, not very different from the Senki, and from the common colick, in other pains of the lower belly, where the cause of the distemper lies in the guts, out of reach both of the needle and Moxa; and in several other diseases, which I here forbear mentioning. It is a powder, to be taken inwardly, and called by the common people, Dsiosei, and in the language of the learned, Wadsusan. It is sold in the village

Menoki, in the province Oomi, sealed up by the inventor, who, by a religious fraud, obtained a privilege for the sole disposal of it. For he gave out, that the ingredients of it, being vegetables, were shewn him by the God Jakusi in a dream, growing upon a neighbouring mountain, which is otherwise famous for many fabulous stories, said to have happened on it, and in the neighbourhood. The good effect people found upon taking; it, soon brought it into repute, and the freat consumption there is of it, enrich'd that whole family, which was formerly very poor, but became afterwards able to build three temples, as publick and lasting monuments of their gratitude to the God, who communicated the secret to them. These temples stand opposite to three shops, where this powder is now made and sold. I brought a Quantity of it with me out of Japan, but found upon trial, that it would not at all agree with my Countrymen. It is bitterer than gall. The preparation of it is kept a secret in the family. However, upon seeing some of the ingredients in a shop, where I bought mine, I took notice, that the bitter sort of Costus, which is called Putsjuk, and is imported into Japan by the Dutch, who bring it from Suratte, was one of the chief; the virtues of this Costus are said to be very considerable, and there is a much greater demand for it in Japan, than for any other exotick drug, excepting only the root of the Sisarum montanum Coraeense, or Ninsin of Dr. Cleyer.

An account of the Moxa, an excellent Caustic of the Chinese and Japanese, with a Scheme shewing what parts of the human body are to be burnt with that Plant in several distempers

There are in Asia three Helicons, that of the Arabs, Bramines and Chinese. Whatever nations inhabit that vast extent of ground, which reaches from Europe to the very extremities of the East, and so far as our Antipodes, have all the arts and sciences flourishing among them, derived from these three chief seats of the Eastern Muses. I forbear enlarging at present upon several things, which might be urged in proof of my assertion, confining myself only to what relates to my own profession. It is not in the least to be wonder'd at, that so many nations, and these so widely differing in their religion, customs, language, and the very nature of the climate, which they inhabit, should have also different principles of the healing art, different remedies, different precepts and methods of cure. The differing Helicons, which gave birth to all the learning of the East, easily account for it. Thus far however they are observed to agree, that being ask'd their opinion about the causes of distempers, they have so frequent a recourse to winds and vapours, that they seem, m imitation of our divine Hippocrates, Lib. de flat, to look upon them as the general causes of almost all diseases incident to human bodies, particularly those which are attended with pain. Upon this principle is grounded their method of cure, and the frequent use of caustics, which they say are the most effectual remedies to discuss and expell all manner of winds and vapours. But then indeed it is a great question with them, what sort of Caustics are the most proper to answer this end, whether fire, or hot irons? To try the joint strength of Vulcan and Mars upon human bodies, they esteem a cruelty, not only needless in itself, and to no purpose, but altogether unbecoming a rational Physician, who can, and ought to have no other intention in the application of Caustics, but to discuss and resolve the viscid matter, which is the cause of the pain and distemper, and afterwards to make room for it to come out. Hence it is, that they are more favourably inclined for a slow and gentle burning, and, in a word, will prefer those Caustics, which are found proper, by vertue of their aperitive salts, to open and dissolve the obstructions, and to draw out the cause of distempers, slowly indeed, but with safety, that, I say, they will preferr them before all the cruel apparatus of other more violent cauteries, which by their sharp and burning vitriolick and cutting quality, miserably corrode and destroy the parts they are applied to. For the same reasons it is, that the ancient Egyptian, Greek and Arabian Physicians, to whom we Europeans are indebted for the invention and many improvements in the Physical art, chose to apply burning mushrooms, or the fiery roots of Struthium and Aristolochia, preferably to hot irons: That some others used hot melted Sulphur; others again spindles of box, dipt in burning hot oil, and applied to the affected part. But it is foreign to my present purpose to enumerate all the various Caustics in use among the ancient Physicians. Whoever hath a mind to be farther informed about this matter, may consult Mercatus, Pr. L. 4. c. i. p, 162, or M. A. Severinus, among the modern writers.[1] My design is to give some account of those Caustics only, which are in use, at this day, in several Asiatick Countries.

---

[1] [Ludovicus mercatus (16th century) and Marcus Aurelius Severinus (1580-1656).]

§ 2.

The Arabians, and those Asiatick nations, which received
their arts and sciences from them, as, for instance, the
Persians, and those of the Great Mogul's subjects, who
embraced the Mahometan faith, so far as I could learn upon
diligent enquiry, never make use of any other Caustic, but
woollen doth dy'd with woad, or what the French call
Cotton Bleu. They take a piece of this blue cloth, wrap it
together, tight and close, into the form of a Cylinder, about
half an inch in diameter, and two inches long. They apply
this Cylinder to the part, and then set fire to the top of it,
letting it glow and burn down insensibly, till it is quite
consumed into ashes. This Caustick is not only extremely
painful, but besides lasts very long, and troubles the patient
sometimes a quarter of an hour, and longer, before it is
burnt out, and the heat over. It is likewise attended with
very bad consequences, frequently corroding and eating
through the flesh, so as to occasion sordid and almost
incurable ulcers, which I know to be true, insomuch, as
during my stay in those Countries, many patients under
these circumstances applied to me for relief. The burning
being over, the Surgeon hath nothing more to do, but to
anoint the part, and when the Eschara, or Crust comes off,
to promote the suppuration. I am apt to believe, that the
extreme and lasting pain, occasioned by these Caustics, and
the great difficulty of curing the ulcers, which too frequently
follow the application thereof, are the reason, why the
inhabitants of these Countries make so little use of them,
for all they are so much commended by their Physicians in
their writing and conversation. I have just now mentioned
the Glastum, or Dyers Woad, and must beg leave to add
something farther upon this subject. The Caustics of the
Arabian Physicians must be of a substance died with the

decoction of this Plant, upon a supposition, that it encreases the Force of the fire, which supposition, they say, is far from being imaginary, but grounded on a continued experience of many centuries. This opinion of the Arabians is also supported by a notion, which very much prevails among the common people in Europe, that burning a piece of cloth dyed blue with dyers-wod, and holding it under the nose of People in Epileptick convulsions, or possessed with the Devil, as some call it, will take off the fit more effectually, than the smoak either of white linnen, or any other stuff whatever. Thus much I can affirm, as matter of fact, that in my own practice in the Indies, I found it very successful in external inflammations, to apply blue bandages and rags, in fomentation and otherwise, instead of common white linnen, to which in the like cases they are certainly preferable.

2. Among the Brahmines, and Indian Heathens

The Brahmines, or Gynmosophistse of the ancient Greek writers, who are the Philosophers, Divines and Physicians, of the Indian Heathens, and all those Pagan nations, which follow their doctrine, do not confine themselves to one single Caustick, like the Arabians, but make use of many, according to the variety of cases and distempers. They say, that the hidden causes of diseases are not all of the same kind, and that their changes are equally various, that consequently the use of one single caustick cannot with any probability be supposed equally successful in all cases, but that such a one must be chosen, as hath been founds by repeated experiments, to agree best with the nature of the distemper, and the constitution of the patient. But what various sorts of Caustiks the Brahmines make use of, and how they ought to be applied, I could not learn, for all I diligently enquired, as indeed it is almost impossible for foreigners, in general, to penetrate into the secrets of these mysterious doctors. The most common Caustick, used in these Countries (for the rest, whatever they be, are applied but seldom) is the pith of the Junci, or rushes, which now in common Scirpus. It is no matter, what sort of rushes it be, provided it be somewhat thicker and larger than the common Scirpus. This pith they dip into Sesamus's-seed-oil, which plant grows in great plenty in their fields, and bum the skin with it after the common manner. I took notice, that the Malayans, Javans and Siamites make use of this pith in burrying their dead, which custom, it is highly probable, obtains also amongst several neighbouring nations.

Advancing still father beyond the Ganges, we shall there meet with another excellent Caustick, referable to all the rest, and very much used by the Chinese and Japanese. These two nations trace up its origin to the remotest antiquity, and pretend that it was known long before the invention of Physick and Surgery, and that consequently the use of it is sufficiently supported by a continued experience of so many ages. This ancient and so much commended Caustick goes by the name of Moxa, not only in China, but in all other Countries, where the learned characters and language of the Chinese are known, as in Japan, Coraea, Qiunam, the Luzon, or Philippine islands, the island of Formosa, and the kingdoms of Tunquin and Cotsijnsina. 'Tis the history of this Caustick, I now propose to give, flattering myself, that the reader will easily excuse, if instead erf the Chinese names, which I am very sensible would be the most acceptable, I insert the Japanese ones, which I did not only for their being easier, but chiefly, because having staid in the Country myself for some time, I was better acquainted with them.

§ 3. Preparation of the Moxa

Moxa is a soft down, or flaxy substance, of a grey or ash-colour, very apt to take fire, though it burns but slowly, and with a moderate heat, there being scarce any sparkling observed, till it is quite consumed into ashes. It is made of the dry leaves of the Artemisia vulgaris latifolia, or common mugwort with broad leaves, which are pluck'd off, when the Plant is very young and tender, and hung out in the open air for a long while. The Japanese say, that it is not at all times equally proper to gather the mugwort for making the Moxa, but that it must be done only on such days, which have been by their Astrologers singled out for this purpose, and have the advantage of a particular benign influence of the Heavens and stars, whereby the virtues of this Plant are greatly increas'd. These days are the first five days of the fifth Japanese month, call'd Gonguatzgonitz by the natives, which according to the Gregorian almanack answer to the beginning of Junee, and sometimes, but seldom, the latter end of May. For, as I have elsewhere observed, the Japanese begin their year with the new-moon, which is next to the spring equinox. The Plant must be gathered early in the morning, before it loses the dew, which fell in the night, and then hung out in the air on the Westside of the house, till it is mil dry. It is afterwards laid up in the garret, and it must be observed, that the older it is, the tenderer and better down may be obtained from it, for which reason some keep it ten years. The fresh and young Mugwort is by the Japanese call'd Tutz, and, when it is full grown, and come to perfection, they call it Jamoggi. And here I cannot forbear taking notice, that it is customary, both in China and Japan, for men to change their names, when they come of age, or have been rais'd to any considerable post. In the like manner different names are frequently given to Plants (not to mention other things) according to their different state of perfection, and differing uses. This variety of names, 'tis true, conveys to our mind a dear and distinct idea of things, as they are at different times, and under different changes, but on the other hand it so multiplies the numbers of words, as to become very troublesome to the memory. The preparation of the Moxa is a matter of no great art or difficulty. In the first place, the leaves are beaten with a pestle into the form of a coarse flax, and then rubb'd with both hands, till they lose the coarser fibres, and harder membranous parts; which being done, there remains only that soft, delicate, homogeneous, and so much commended down, which nature bestowed on the young Mugwort preferably to other plants.

§ 4.

The burning of the Moxa hath nothing in the least to terrify people, and to deter them from going through the operation. It bums so slowly, that scarce any sparkling can be discern'd, and it might be doubted, whether it burns at all, were it not for a thin scarce visible smoak arising from it, which however is not at all disagreeable to the smell. The pain is not very considerable, and falls far short of that which is occasioned by other Causticks, or actual Cauteries. Those Cones indeed, which the Japanese call Kawakiri, that is, Skin-Cutters, are something more painful, being the first two or three tents successively applied to the skin. 'Tis from these Cones that the Japanese call the new taxes, laid on them by their Princes and Governors, Kawakiri, because they say they are very hard to be bore at first, but become much easier in time. I have seen many times the very boys suffer themselves to be burnt in several parts of their body, without shewing the least sense of pain: For they bum indifferently, and without regard, old and young, rich and poor, male and female; only women big with child are spared, if they have not been burnt before. The intent or burning with the Moxa is either to prevent or to cure diseases. But it is more particularly recommended by their Physicians as a preventive medicine, for which reason they advise the healthy, more than sick people, to make use of it. This practice of theirs they ground upon the following principle, that by the very same virtue, whereby it dispells and cures present distempers, it must of necessity destroy the seeds of those to come, and by this means prevent them. Hence it is, that in these extremities of the East, all persons, who have any regard for their health, cause themselves to be burnt once every six months. This custom is so thoroughly and so religiously observed in Japan, that even those unhappy persons, who are condenm'd to perpetual imprisonment, are not deprived of this benefit, but are taken out of their dungeons once in six months, in order to be burnt with the Moxa. The burning with the Moxa, by way of prevention, requires but a few tents, and those very small ones, but if it be intended to cure a distemper, there must be more, and larger, particularly if the cause of the distemper lies deep, and is consequently so much the more difficult to be removed.

If you ask either the Chinese or Japanese, in what distempers it be proper to bum with the Moxa, they return the following answer. That it is proper in all those distempers, where an occult vapour, and which lies, as it were, imprisoned within the body, occasions a dissolution of the solids, and a sense of pain, and hinders the affected part from duly performing its functions. Considering things in this view, there is scarce a distemper, of all that infinite number, incident to human bodies, but the Japanese and Chinese Physicians will advise their Patient to be burnt with the Moxa for it, which quickly, as they pretend, and in a very short time, destroys and removes Its cause. This Caustick is not unknown to those black Asiatick nations, which inhabit the torrid Zone. They learnt it from their neighbours, and it is not long ago that its use was introduced among them, with that difference only, that they apply much larger tents, or cones, than either the Chinese or Japanese, of whom they had it, in proportion as the distemper is difficult and dangerous, or as its cause lies deep in the body. Even the Dutch in the Indies have lately experienced, what a good effect may be expected from burning with the Moxa in arthritick, gouty, and rheumatick distempers. This Caustick breaks the force of the saline and tartarous particles, which the too plentiful use of Rhenish wines leaves in the blood, and which being fix'd about the joints, and particularly irritating that sensible membrane, which encompasses the bones, are the cause of gouty paroxysms. It dissolves the stagnating lymph, which being gather'd about the articulations, occasions Rheumatick and Arthritick pains, provided a larger cone or tent be applied for either of these purposes, and provided it be applied in time, before the morbid matter be accumulated so far as to break and lacerate the capillary vessels, to tear the membranes and muscles, in which it is lodged, and thereby to occasion those tumours and impostumations, which are frequently the consequences of these dangerous distempers, and which will then yield no farther to any emollient or dissolvent medicine whatever. However, it may not be amiss to observe, that although in the hot Asiatick

Countries the use of this Caustick hath been found upon experience very successful in the above-mention'd distempers, yet the like success cannot be reasonably expected from its application in our colder European climates. In hot Countries the perspiration is stronger, the fluids thinner, the pores wider, the muscles and membranes more relaxed. Sometimes also, by the application of this Caustick, the pain will be only removed, and not entirely taken off. The force of the saline particles will be broke in those parts, which are burnt by the Moxa, and sometimes perhaps it will penetrate so deep as to burst and tear the periostium. This will doubtless take off the sense of pain in these very parts, but be no hindrance to its shifting to others. The Brahmines indeed go farther, and confidently assure their patients, that the pain, being once removed, will never return, if they do but abstain from eating of flesh, and from strong fermented inebriating liquors, such as wine, beer, and the like. These, they say, breed -new crudities, which, when they come into the blood, will fell down again upon the legs, and there lay a new foundation for gouty paroxysms. Bushofius, a Minister of the Gospel at Batavia in the Indies, went too far, when he recommended the Moxa to his Countrymen in Europe, as an infallible remedy for the gout. I have reason to apprehend, that many a patient in Germany found himself disappointed in his expectation: This is what the learned Dr. Valentini, a German Physician, and Member of the Academy of Sciences founded by the late Emperor Leopold, complain'd of, and not without reason, in a printed letter of his to Dr. Cleyer, to whom it was delivered in my presence. The neighbouring black Asiatick nations make more use of the Moxa, than the Chinese and Japanese themselves, in Epileptic fits, and all Chronical distempers of the head. Their way is to bum a good quantity of it all along the Sutura Coronalis, which sometimes hath been attended with so good a success, that some patients recovered, who had been given over by the Physicians.

§ 5. Places to be burnt with the Moxa

The Chinese and Japanese Physicians widely differ in their opinions concerning the parts of the human body, which it is proper to burn with the Moxa, in order either to cure, or to prevent particular distempers. And although superstition and self-conceit have a very considerable share in their reasonings, yet they all plead either their own experience, or that of their master, for what they assert. If their different opinions were to be brought together, I believe, that in some distempers there would be scarce any one part of the human body left, but what some of them would single out as the most proper to be burnt with success. The common people seldom recede from the common places and rules, handed down to them from remotest antiquity, and represented, for the benefit of the publick, in particular printed schemes. They are still more superstitious about choosing the proper tune, when particular parts of the human body ought to be burnt in particular diseases: And here great regard is had to the situation and influence of the Constellations of the Heavens, for it is agreed on all hands, that even when they are come to a resolution, what parts it is proper to bum, yet the operation ought not to be performed on an ill day, and in an ill hour, when, according to their way of reasoning, the less favourable influence of the Stars gives room to apprehend an ill success. In this again their judgment and opinions are so various, that if there was any attention given to what every one in particular thinks and advises, it would be scarce possible to find any good day or hour at all. What they chiefly aim at in choosing the proper places for burning with the Moxa, is to find out such as are the most conveniently seated, either to draw out the vapours, which are the supposed cause of the distemper, or to remove them from the affected part. These they all pretend to be well known to them by the observations of their ancestors, and by their own experience. No part of the human body suffers so much by this Caustick, as the back side, all along the Spina Dorsi, on both sides quite down to the loins. I found the backs of the Japanese (and this is likely to be the case of all other Asiatick nations, that make use of the Moxa) of both sexes so full of scars and marks of former exulcerations, that one would imagine they had undergone a most severe whipping. But to whatever degree they be disfigured by the Moxa in this and other parts of their body, their beauty is, according to their notions, not in the least lessened thereby. And as to the back in particular, it is a very easy matter for the Japanese to uncover it, and they do it very frequently when they go even about a slight work, letting their gowns, which are tied about their girdle, fall down behind their back, lest they should be spotted with their sweat, they wearing no shirts, by which means their wounds and sores, in both sexes, are kid open to view.

§. 6.

I come now to the operation itself, which requires no great nicety or skill. A small quantity of Moxa is rolled or twisted, between the thumb and fore-finger, into the form of a Cone, almost an inch high, and something less broad at the bottom. This Cone is put on the part which is to be burnt. Some wet the bottom a little with spittle to make it stick to

the skin. This done, they put fire to the top with a thin
burning splinter, which the Japanese call Senki. The Cone
being consumed, which is done in a very short time,
another, if needful, is applied to the same part, and burnt as
before. This is repeated as often as the Patient desires, or
the Operator directs, or the case seems to require. The
Surgeons, whose business it is to perform this operation,
are call'd by the Japanese Tensasi, that is, feeling people, or,
according to the literal sense of the word, people that
penetrate with the touch, because, before the operation,
they always feel about, and examine the part, which the
Caustick is to be applied to. The little rods, or candles, which
they make use of to put fire to the Caustick, are the very
same which the Heathen Priests bum in the temples before
their idols, and whereby they measure the hours of
devotion, in imitation, as it were, of the fires, which it is
customary to make in camps, to indicate and to measure
the time for watching. They bum but slowly, and have a
very fragrant strong scent. They are made of the slimy bark
of the Taab tree, as they call it, or Taabnoki, that is, Laurus
Japonica sylvestris, wild Japanese bay-tree, one of the
tallest and largest trees growing in the Country. This bark is
reduced into a powder, and mix'd with Aloe wood, or its
resinous and dearest part, call'd Calamback, and with other
sweet-scented species, according to every one's fancy, all
reduced into a powder. These powders are mix'd with water
to the consistence of an Electuary, or thick pulp, which must
first undergo a sufficient kneading, and being then put into
a bason with many small round holes at the bottom, and
weights being laid upon it, there are squeezed out through
these holes long round pieces, or rods, scarce thicker than a
straw, which being taken off, are laid on lathes and dried in
the shade, and afterwards sold in shops for burning candles,
and for the use above-mention'd, by bundles wrapt up in

paper. These Senki candles however are not so absolutely necessary for the operation, but that they may be rank'd rather among the more elegant and less useful Apparatus of Surgeons. Any common splinter, or straw, will answer the end full as well, and these are what the common people make use of. The main art lies in the knowledge of the parts, which it is proper to bum in particular distempers. The chief intention of burning is, to draw out the humours and vapours, which lying concealed in the body, prove the cause of the sickness. And although, upon this supposition, one would reasonably imagine that place to be the most proper which is the nearest to the affected part, yet the operators frequently choose such others, as are not only very remote from it, but would be found, upon an Anatomical inquiry, to have scarce any communication with it, no more than by the common integuments. As strange as that Polish nobleman thought it, to have a clyster ordered him, when he complain'd of a pain in his head, so surprizing will the effects of this Caustick appear to foreigners, when applied to places which seem too remote from the affected part, to suppose any communication with it. A few instances will serve to explain this. In Indigestion, and sickness of the stomach, and loss of appetite, they apply the Caustick to the shoulders. In pleuritick cases they burn the Vertebrae of the back, and in the tooch-ach the adductor Muscle of the thumb, on that side where the pain is: and so on. I am sensible, that the most skilful Anatomist would be at a loss to find out any particular correspondence of these remote and differing parts with one another.

§ 7.

There are several things required, and many particular rules to be observed, in the application of this Caustick, with regard chiefly to the place which is the most proper to be burnt, to the time, when the operation is to be performed, to the number of Cones, which must be applied successively, to the situation of the Patient, when under the operation, to the proper diet to be undergone both before and after, and other the like circumstances. The following are the chief and most general rules. Tendons, Arteries and Veins must be avoided with all possible care, in order to which the operator must not only call to help his eyes, in a careful examination of the parts, but make use also of his fingers, and feel whereabouts they lie. Whatever situation the Patient was in, when the properest place for the application of the Caustick was examined and determined, in that same he must remain, whilst the operation is performed, whether he was sitting or standing. He that is to be burnt, must sit on the ground cross-leg'd, after the fashion of the Eastern nations, holding the palms of his hands to his cheeks, that posture being the nearest to that in the mother's womb, and thought the most proper to shew the situation and interstices of the muscles. Those that are to be burnt in the legs, must sit on a stool or chair, holding their legs down into a tub of warm water, because, they say, that in these parts, which are so remote from the fountain of heat, the perspiration must be promoted by art. Those persons, who are of a tender sickly constitution, must not have more than three Causticks applied at a time, to any part of their body whatever. To strong people ten, twenty and more, must be order'd, according to the nature of the distemper. There are no certain rules to go by, as to the number of Cones, which must be burnt on any part successively, or whether the same must be applied alternatively, this depending in a great measure upon the Patient's patience, and the operator's pleasure. The day after the operation, and for some following days, the operator examines and dresses the part. If he finds it dry and not suppurated, he looks upon it as a very bad sign, and a proof that nature is scarce strong enough to throw out the morbifick matter. In this case he endeavours to promote the suppuration, by applying pounded onions. Thus far what I could learn concerning the Moxa, by conversing with the Surgeons of the Country, and those persons, who make it more particularly their business to burn people with it.

As to the more particular rules of this burning art, they have tables printed in Chinese and Japanese characters, of which I here present the Reader with one, which I endeavoured to explain and translate, so well as the nature of the Chinese verse, wherein it is wrote, and the principles of their Philosophy would admit of. I have likewise added two Schemes, (Figure 7) being two different views of the human body, wherein is shewn, what parts are proper to be burnt in certain distempers, with the particular names of these parts. They are sold in booksellers shops, and by mountebanks, who cry them up in the streets and publick places, to allure the common people to buy, for a trifle, all the rules and precepts of an art, which they are ignorant of. The text, as I found it in the Japanese orginal, is printed in Italick characters, and the few notes, which I was able to add to explain the same, in Roman, enclosed within two hooks.

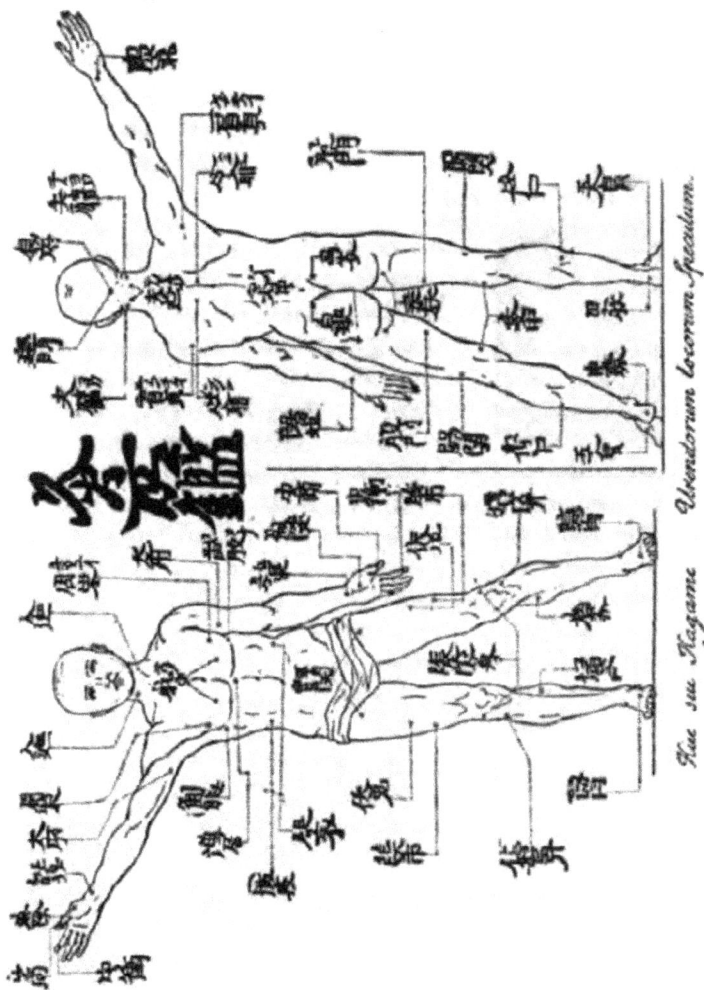

Figure 7. Two schemes shewing what parts of the human body are to be burnt with the Moxa in several distempers

Kiusiu Kagami - A Treatise (in the literal sense a Looking-glass) shewing what parts of the human body are to be burnt with the Moxa[1]

Chap. I.

Shews the method of burning deliver'd in verse in certain propositions, whereby this whole art is discover'd to the world.

1. *In the head-ach, swimming of the head, fainting fits, in the* Dseoki, (Dseoki is a particular kind of an inflammation in the face, occasioned by a scorbutick disposition of the body, which is very common in this Country. Persons, who labour under it, are frequently affected with swellings in their faces, and sometimes the whole head, attended with an almost intolerable sense of heat and burning, and this very often from slight causes, as from bathing, and excesses in drinking, and exercises. This swelling is often followed by an inflammation of the eyes.) *in a dimness of the eyes, occasioned by a too frequent attack of the Dseoki, in pains of the shoulder after head-ach, in asthma's and streightness of breath, it is proper to bun that part of the human body, with is call'd* KOKO.

2. *In distempers of the Children, particularly swellings of the belly, loosenesses, loss of appetite, in the itch and exulcertion of the noses, as also in shortness of sights the region of the* SIUITZ, (or eleventh vertebra) *must be burnt*

---

[1] [灸所鑑, kyūsho kagam; The 'Moxa Mirror'. See especially W. Michel's works on Kaemfer and Japan for an analysis of this section.]

on both sides with fifteen or sixteen cones, leaving one SUN *and a half's distance {about two or three inches) between the two places, which they are to be applied to. Remark 1. Siuitz,* or the eleventh, is so called from its being the eleventh vertebra in number, computing from the fourth vertebra of the neck, that being the most apparent of any, when the head is bowed down forwards towards the breast. The same rule must be observed with regard to all the other vertebrae, whereof the number only is mentioned. *Remark 2.* Sun is properly speaking a measure, whereby they measure the length or things. They are of two different sizes, the longer is made use of by merchants, the shorter by builders, and Workmen. The *Sun,* as it is above mentioned, with regard to the method of burning with the Moxa, must be understood of neither of these, but its length taken from the second joint of the middle-finger of that very person on whom the operation is to be performed, as bearing the most accurate proportion to other parts of the same body.

3. *In the Sakf* (a chronical and intermitting kind of a colick,) *in the Senki,* (or that colick, which is endemial to this Country[1]) *and in the Subakf,* (or gripings of the guts occasioned by worms) *you must bum on both sides of the navel at two Suns distance. This place is called* TENSU.

4. *In the obstruction of the menses, and influxes; in whites, in piles, and the exulceration of the haemorrhoids, and in the Tekagami,* (an intermitting sort of a cold, attended with pain and heaviness in the head) *you must bum the place* KISOO *or* KITZ, *on both sides with five cones. To find out this place, you must measure from the navel streight down four*

---

[1] Amply treated in Num. 3 of Kaempfer, *History of Japan,* 1727.

Suns, then sidewards at right angles four Suns on each side, so that there be eight Suns distance between the two places to be burnt.

5. *In a difficult delivery you must bum three cones on the extremity of the little finger of the right foot. This will give instant relief and promote the delivery.*

6. *In want of milk in nurses, five cones must be burnt between the two breasts in the middle.*

7. *In arthritick pains and rheumatisms, in pains of the legs, as also in strangury, or retention of urine, you must bum about eleven cones, on the thighs about three inches above the knees, (or on the place for issues.)*

8. *In swellings and pain of the belly, in pain at the heart from a quotidian fever, in pain of the stomach, and loss of appetite y you must bum six cones above the navel. The place, which you are to burn, must be four Suns distant the navel, in a streight line upwards.*

9. *In pain of the hips and knees, for weakness of the legs a particular, and of all members of the body in general, you must burn the place, called JUSI. (Jusi is that place on the thighs, which one may reach with the extremity of his middle-finger, holding his hands streight downwards in a natural situation.*

10. *Those, who have a hardness and swelling in the Hypochondria, as also those who have frequent shiverings, or relapses of putrid fevers, must be burnt in the place called* SEOMON. (*Seomon* is just beneath the last false rib on each side. The burning of this place is extream painful. I should have thought it more proper to write it *Schomon*, or *Siomon*, but hearing the Japanese pronounce it themselves, I found that they make a short *e* of it.)

11. *In claps you must bum in the middle of the place called* JOKOMON. (*Jokomon* is above the privities in the middle between them and the navel.)

12. *Those persons who are subject to colds, bleeding at the nose, or swimming of the head, will find great benefit, if they cause from fifty to an hundred cones to be burnt* (successively) *in the place, called* TUUMON. (*Tuumon* is the region of the Os sacrum.)

13. *Those who are troubled with tumours and ulcers in the anus, must have one cone burnt three suns from the extremity of the Os Coccygis*: (The burning of this place is attended also with a very great, and almost intolerable pain.)

14. *In the procidentia ani, the Os Coccygis itself must be burnt.*

Chap. II

Nindsin, (the spirit of the Stars) *lodges in the spring about the ninth vertebra, in the summer about the fifth vertebra, in autumn about the third, and in winter about the fourteenth, and near both hips: For this reason care must be taken not to bum any of these places, at the times above-mentioned.*

2. *Upon the turning of the four seasons of the year, you must avoid burning either the place, called Seomon, or the fourteenth vertebra, because instead of being beneficial, it would rather prove hurtful, and encrease the distemper.*

3. *You must entirely abstain from burning in rainy, wet, or too hot weather, and on a cold day.*

4. *You must not lie with your wives three days before, and seven days after the burning.*

5. *Angry, passionate people must not be burnt, before their passion is calmed. Weary people, and who are just come from their work, must not be burnt, till they have rested themselves. The same rule is to be observed, as to hungry people, or such as have eat too much.*

6. *People must abstain from drinking of Saki* (a spirituous fermented liquor, brewed out of rice) *before they are burnt,*

37

but after the operation hath been performed, it is not only safe but advisable to do it, because it promotes the circulation of the spirits and blood. (The Japanese knew long ago, that the fluids circulate in our body, but how, and after what manner the circulation is performed, they are still ignorant of.)

7. *Great care must be taken not to go into a bath of sweet watery for three days after the operation.* (The Japanese are very great lovers of bathing, and use it every day. I believe that this is the reason why the pox spreads so much less, than it would be otherwise like to do in so populous a Country.)

8. *Medicines should be given to cure the distempers incident to our body, and the burning with the Moxa should be ordered to preserve us from them. For this reason even those, who are otherwise in a good state of healthy should be burnt twice a year, once in the second month* (March) *and once in the eighth* (September.) (The proper days for burning, and which are favoured by the influence of the Stars, are set down in their almanacks.)

9. *You must feel the pulse before you burn: If it be too Quick, you must act prudently, because that shews that your patient hath got a cold.*

10. *The places to be burnt, must be measured by SAKU and SUNS. The length of the Sun must be determined from the*

middle joint of the middle-finger, in men in the left and a women in the right hand.

Chap. III.

Women who would have done breedings must have Am cones burnt on the navel.

Chap. IV.

Women that would be glad to have children, must have eleven cones burnt on the side of the twenty-first vertebra.

2. A treatise on acupuncturation – Churchill, 1821

A treatise

on

acupuncturation;

being

a description of a surgical operation originally peculiar to the Japonese and Chinese, and by them denominated

zin-king,

now introduced into European practice,

with

directions for its performance,

and

cases illustrating its success.

by

James Morss Churchill,

Member of the Royal College of Surgeons in London.

A TREATISE

ON

ACUPUNCTURATION, &c.

Dedicated, by permission,

To

Astley Cooper, Esq. F. R. S.

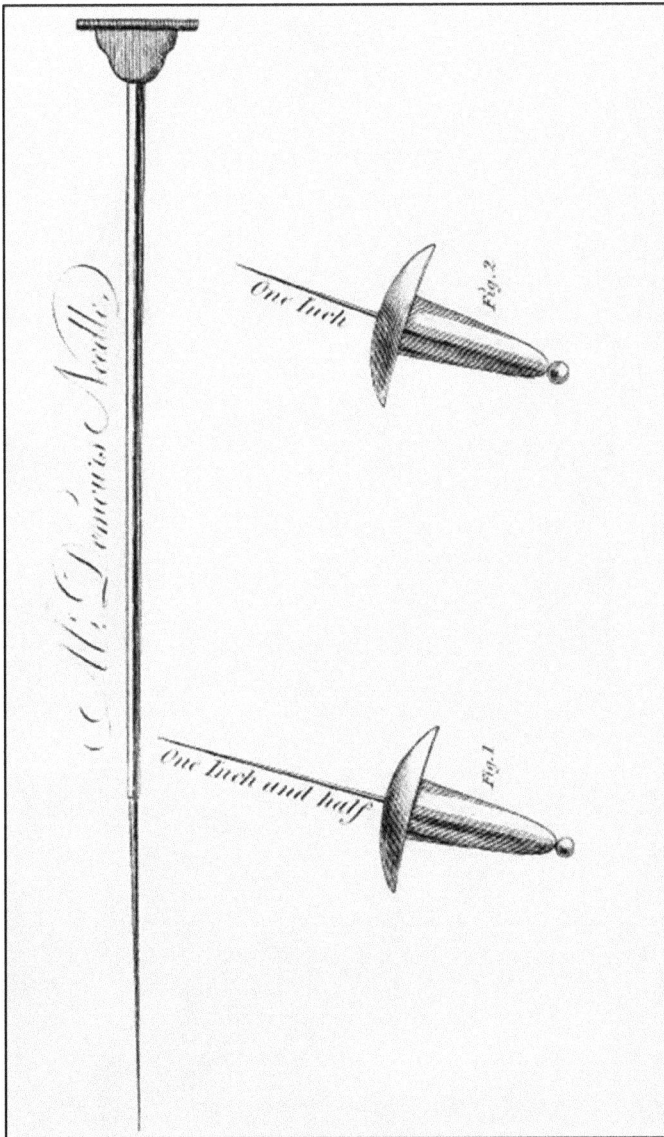

Figure 8 - Acupuncturation needles

TO

ASTLEY COOPER, ESQ.

THE STEADY FRIEND AND PATRON OP HUMBLE MERIT, THE
AUTHOR RESPECTFULLY INSCRIBES

THIS LITTLE TREATISE;

LESS FROM PRESUMPTION OF ITS DESERVING HIS
APPROBATION,

THAN

AS A MARK OF RESPECT

FOR SPLENDID ACQUIREMENTS,

GRATITUDE,

TOWARDS A GREAT MASTER.

Works by James Morss Churchill

*A Treatise on Acupuncturation : Being a Description of a
    Surgical Operation Originally Peculiar to the Japonese
    and Chinese, and by Them Denominated Zin-King, Now
    Introduced into European Practice, with Directions for
    Its Performance, and Cases Illustrating Its Success.*
    London: Simpkin and all, 1821.
    http://archive.org/details/treatiseonacupun00chur

Stephenson, John, and James Morss Churchill. *Medical
    Botany : Or, Illustrations and Descriptions of the
    Medicinal Plants of the London, Edinburgh and Dublin
    Pharmacopoeias; Comprising a Popular and Scientific
    Account of All Those Poisonous Vegetables That Are
    Indigenous to Great Britain*. London: Churchill, 1831.
    Vol. 1. http://archive.org/details/b24923436_0001
    Vol. 2. http://archive.org/details/b24923436_0002
    Vol. 3. http://archive.org/details/b24923436_0003
    Vol. 4. http://archive.org/details/b24923436_0004

Preliminary Remarks

IF the medical profession merit the reproach, of being easily
deluded into an admiration of novelty, then I need use no
apology for introducing the following pages to notice, nor
will my subject stand in need of prefatory allurements to
obtain attention; but if on the other hand, a rational theory,
built on sound logical reasoning, be the only evidence to
which any value can be attached, then will my efforts have
been unavailing and fruitless. Under the impression,
however, that there exists a desire for speculation and
discovery on the one hand, regulated and qualified by a

moderate and proper degree of scepticism on the other, I shall presume a medium of the two extremes, and proceed without apology or preface to my subject, trusting, that the interesting facts which I have to relate, will elicit such attention and investigation, as will kindle a desire in some men, at least, to become acquainted with a process, which appears to rival the most successful operations for the relief of human sufferings.

I should not have taken the tales which are told of the wonderful cures effected by this operation amongst the original founders of it, as sufficient authority for recommending it, nor would I admit the fables which are promulgated by these people, as evidence of its efficacy, had not this efficacy been witnessed by European spectators on its native soil, and at length experienced in our hemisphere; and even, latterly, in our own country.

The operation of acupuncturation has been seen by so few Europeans, that our books have made us acquainted with little more than its name. It is of Asiatic origin, and China and Japan peculiarly claim it as their own. A writer in the year 1802, mentions a discovery of its having been practised by the natives of America, and refers to Dampier's voyages tor an account of it; but I have in vain followed Capt. Dampier's relation of his adventures, in crossing from the South to the North Sea, over the Isthmus of Darien, for any account of the operation, for he does not so much as name it. He speaks of a work intended to be published by his surgeon, Mr. Lionel Wafer, who accompanied the expedition, and to which he refers his readers for an account of the manners and customs of the interior of the country. Mr. Wafer was detained, from an accident, a considerable time amongst the Darien Indians, and did, on his return to England, publish this book, which I have therefore been at the trouble of perusing, but do not learn from it, that the operation of acupuncturation was practised in that part of America: it is true, Mr. Wafer describes a method of blood-letting employed by the natives, which is somewhat correspondent to acupuncturation, but both the intention and the effect are widely different. This operation is effected in the following manner: the patient is taken to a river, and seated upon a stone in the middle of it. A native, dexterous in the use of the bow, now shoots a number of small arrows into various parts of the body. These arrows are prepared purposely for this operation, and are so constructed, that they cannot penetrate beyond the skin, the veins of which, opened by the puncturation, furnish numerous streams of blood, which flow down the body of the patient. If this be the operation which has given rise to the idea, that acupuncturation is practised by the American natives, the conclusion is evidently erroneous, as it is simply a method of blood-letting, and is generally resorted to for the cure of fever. Now, acupuncturation has no reference whatever to bleeding, and it is rare, that even a drop of blood follows either the introduction or withdrawing of the needle; nor does it appear, that the Chinese and Japonese, with whom it originated, intended it as a method of abstracting blood, which is proved, not only by the consequences of the operation, but by the manner in which it is performed, and the nature of the diseases to which it is applied. If it could have been established, that the natives of the American Isthmus were acquainted with it, it would have been a curious, as well as an interesting enquiry, to ascertain whence they derived it.

It is a little strange, that the surprising efficacy, of which so much has been boasted by its eastern professors, and the safety, at least, with which acupuncturation may be performed, having been so fully demonstrated; it is strange

I repeat, that it has not met with an earlier encouragement amongst us. It is probable, that the hyperbole in which it has been related, has induced the sober minds of our Northern soil, to treat these relations as the fictions of Eastern imagination, and to reject them without examination, as fables calculated only for amusement. There have not, however, been wanting sensible minds, and men of talent and reputation, to recommend this operation; and the names of Ten-Rhyne, Bidloo, Koempfer, and Vicq-d'Azyr, stand conspicuous on the list of those who speak in its favour; but still, neither of them had undertaken to put its merits to the test, by actual experiment. Several practitioners in France, however, have now taken up this neglected operation, and their report verifies the praises which have been bestowed by others upon it. My attention was lately directed to it by my friend Mr. Scott, of Westminster, who, as far as my knowledge goes, was the first who performed it in England, and some successful cases which I witnessed in his practice, assured me of its efficacy, and led me to its adoption. The success of my own subsequent practice, warrants a recommendation of it, in almost any terms I could give it; but I shall content myself in laying before my readers, the opinion and experience of some physicians of eminence, accompanied by a relation of some cases of my own, where the benefit of the operation has been decidedly successful; upon a better foundation than which it cannot at present rest for public examination; it remains for the medical profession to ascertain its claims to attention by the test of experience, and having undergone the ordeal of experimental enquiry, it will, I have no doubt, so fully develope its merit, as to obtain a conspicuous rank in medical estimation, as a valuable curative measure.

Acupuncturation

The method of performing the operation of acupuncturation is simple and easy, requiring neither practice to give dexterity, nor adroitness that it may be done with propriety. Anatomical knowledge of the human body is, however, necessary; as an imprudent application of it, by an operator ignorant of the structure of the part into which he introduces his needle, might be productive of bad consequences. To a surgeon, however, properly qualified, (and no other ought to perform this or any operation) no danger can arise; as the cautions are but few, and no risk is incurred, if they are attended to. It is only necessary that the operator, in introducing the needle, should avoid the course of large vessels, of nervous trunks, and of the tendons of muscles. It is not, however, proved, that the latter sustain injury from the puncture of the needle; but it is as well to avoid the possibility of mischief, by such a cautious mode of introducing the instrument, as shall be divested of risk. I cannot better familiarize my subject to the reader, than by a sketch of it in its native state; and as an excellent description of the operation, as performed by the Japonese natives, is given in the ninth volume of the "Modern part of an Universal History, from the Earliest Account of Time,"[1] I shall extract it, as containing all that is known of its original practice.

"The place made choice of for the puncture, is commonly at a middle distance between the navel and the pit of the stomach, but often as much nearer to, or farther from either as the operator, after a due scrutiny, thinks most

---

[1] [Sale, *An Universal History, from the Earliest Account of Time*, 1759, 30:40-42.]

proper; and in this, and the judging rightly how deep the needle must be thrust below the skin, so as to reach the seat of the morbific matter, and giving it a proper vent, consists the main skill of the artist, and the success of the operation is said to depend. Each row hath its particular name, which carries with it a kind of direction, with regard to the depth of each puncture, and the distance of the holes from each other, which last, seldom exceeds half an inch in grown persons, in the perpendicular rows, though something more in those which are made across the body, thus, ⦂⦂⦂

The needles which perform the operation are made, as was hinted at first, either of the finest gold, or silver, and without the least dross or alloy. They must be exquisitely slender, finely polished, and carry a curious point, and with some degree of hardness, which is given by the maker by tempering-, and not by any mixture, in order to facilitate their entrance, and penetrating the skin. But, though the country abounds with expert artists, able to make them in the highest perfection, yet none are allowed, but such as are licensed by the emperor.

"These needles are of two sorts with respect to their structure, as well as materials; the one, either of gold or silver indifferently, and about four inches long, very slender, and ending in a sharp point, and have at the other end a small twisted handle, which serves to turn them round with the extremity of the middle finger and thumb, in order to sink them into the flesh with greater ease and safety; the other is chiefly of silver, and much like the first in length and shape, but exceedingly small towards the point, with a short thick handle, channelled for the same end of turning them about, and to prevent their going in too deep; and for the same reason, some of them are cased in a kind of copper tube, of the bigness of a goose quill, which serves as a sort of guage, and lets the point in, just so far as the operator hath determined it. The best sort of needles are carefully kept in a case made of bull's horn, lined with some soft downy stuff. This case is shaped somewhat like a hammer, having on the striking side a piece of lead, to give it a sufficient weight, and on the outside a compressed round piece of leather to prevent a recoil, and with this they strike the needle through the thickness of the skin; after which they keep turning the handle about with the hand, till it is sunk to the depth they design it, that is, till it is thought to have reached the seat of the morbific virus, which in grown persons is seldom less than half, or more than a whole inch: this done, he, draws it out, and compresses the part, in order to force the morbific vapour or spirit out.

"The directions and nice rules for the performing of this curious operation are many, and require great skill and attention in the operator; and when duly performed, may be of excellent use, not only against the excruciating distemper, called Senki, but against many other topical ones, which are most commonly cured by the Indian Moxa, and other caustics. On the other hand, these last are often tried against the distemper above mentioned, by applying the caustic to the belly, on each side of the navel, and about two inches from it, but mostly without any success, it being very unlikely that such an application should reach the seat of the distemper; whereas, the benefit which has accrued from the *acupuncture*, in that one disease, hath encouraged others to apply it indifferently to other parts of the body, where the moxa is used, and by a due care and precaution not to prick any nerves, tendons, or other considerable blood vessels, have cured their patients by it, without putting them to the excruciating- torture which attends that of the Moxa, or other caustics."

[There is still another method of curing that and other violent disorders in the abdomen, and lower belly which is still in vogue among the Japanese, though nothing so effectual as the acupuncture: it is a powder taken inwardly, which is only sold in the village of Menoki, in the province of Oumi, sealed up with the arms of the inventor, who, by a pious fraud, obtained the sole privilege of making and vending it. This person, at first very poor, gave out, that the god Jakusi had revealed it to him in a dream, and shewed him the plant growing in a neighbouring mountain, famous among them for many other fabulous stories said to have happened upon or in the neighbourhood of it. The good effects which this remedy produced soon brought it into repute; and the great consumption of it enriched him to such a degree, as to enable him to build a temple to the god above-mentioned; since which, his family, increasing still in wealth, have added two more, as so many monuments of their gratitude to him. Over against each of those three grand structures stands a shop, in which the said powder is made and sold. Our author bought a quantity of them; but upon trial of them, did not find them at all agreeable to his constitution, and of a most distasteful bitter, which he supposes, from some which he saw in the shop, to be the Costus, which is brought thither by the Dutch, in greater quantities than any other exotic, from Surat. However, the powder is chiefly in vogue among the common people in the cholicky distempers above-mentioned; whilst the better sort have recourse to the outward operation of acupuncture, which we have been describing.[1] This, however, doth not hinder the surgeons from using likewise the other method of cauterising; and in some cases, as rheumatism or gout, raise a blister on some nerve with a

---

[1] {Vid. Kaempfer, Append. to Hist. of Japan, p. 29, & seq.}

little powder of mugwort, Moxa, or other herb, and some cotton set on fire.[1]]

From the little we have learned of the practice of this operation amongst the Asiatics, it would seem, that it was chiefly diseases of the abdominal cavity and viscera, which afforded opportunities for its performance, such as Colic, Tympany, &c. It is not in such diseases, however, that I have any experience of its use, but it is questionable, whether it might not be beneficial, particularly in the latter, and I would beg to recommend it as a matter of interesting experiment, to be tried in this malady; such an opportunity, should it fall in my own practice, I shall take advantage of.

The Indians, however, do not confine their practice of Acupuncturation (or Zin-king, as they call it) to diseases of this kind. They puncture the head in all cases of Cephalalgia, in Comatose affections, Ophthalmia, &c. They puncture the chest, back, and abdomen, not only to relieve pain of those parts, but as a cure for Dysentery, Anorexia, Hysteria, Cholera Morbus, Iliac Passion, &c. Local diseases of the muscular and fibrous structures of the body, also often afford them occasions for its performance; and it is for diseases of this class only, that I have hitherto practised it, and for which I would expressly recommend it.

Neither sufficient time has elapsed, nor a proper selection of cases been made since this operation has been known to me, to have afforded me, either a large number of experiments, or a great variety of diseases on which to try the effects of it: it is true I have employed it on some few, and I have it in contemplation to encrease the list, by giving

---

[1] {Ib. ibid. Caron, Varen, &c.}

my experiments a wider range, but at present I should not be doing justice to my subject, to form conclusions on such imperfect evidence; I shall therefore confine myself, merely to the description of the good effects, which I have witnessed in diseases of a rheumatic character, and in those injuries of the fibrous structures of the body, which are often observed to arise, (particularly in labouring persons) from violent exertion. This circumstance must be ever in view, and if it be not fully impressed on the mind, I doubt not but many who may be induced to try the effect of the operation, may be disappointed in it; viz. that acupuncturation does no good, nor does it produce even a temporary alleviation, when the disease for which it is used, is of an inflammatory character. This distinction seems to have regulated the practice of those, who have experimented on the subject, and to have decided them in their selection of cases for the operation. Mr. Berlioz, of Paris, has practised it extensively, and has recently published an account of the success which it has had in his hands.[1] He says,

> "The eulogia given to acupuncturation by Koempfer and Ten-Rhyne, are just and merited. We have reason to feel surprized, that although an age or more has elapsed, since this curative measure has been known in Europe, no physician has made trial of its efficacy. The practice of the operation is attended with but little pain, and the success of it is so prompt, that the disease is alleviated or entirely ceases, as soon as the needle has been introduced the depth of a few lines; most frequently, however, the pain is not removed by the first introduction of the instrument, and it is not

[1] L. V. J. Berlioz, D.M., *Mémoires sur les maladies chroniques, les évacuations sanguines et l'acupuncture*, Paris, 1816."

until after the use of it for a second, third, or fourth time, that the cure is completed. Simple nervous affections, especially demonstrate how much acupuncturation merits the attention of physicians, for there are but few remedies possessed of such prompt activity, and which produce such wonderful effects.

"But acupuncturation does not appertain in any respect to sanguineous evacuations,[1] it can only contribute sometimes to establish the indications for them. This operation is not indeed followed by any success, when the disease depends upon sanguineous turgescence and inflammation.

"In contrary circumstances, Acupuncturation, by dissipating the symptoms, demonstrates, that disorder of the nervous system only had given rise to them."

The only cases of Rheumatism in which I have been successful with the operation, have been of the Rheumatalgic form, or that which is divested of external inflammation; characterised by pain upon motion, stiffness and coldness of the part; the disease having a disposition to change its place; is aggravated by atmospheric changes, and relieved often by stimulant Diaphoretics, Narcotics and external warmth: but I have yet met with success in some cases where the intensity of the pain would have led me to

---

[1] Dr. Haime, whose practice will be presently noticed, observes, "Lorsque l'aiguille a été introduite avec les précautions requises, il n'y a pas émission de la plus petite gouttelette de sang. À ce sujet, le docteur Fréteau est du même avis que M. Berlioz, puis qu'il dit, dans son Traité des émissions sanguines, que l'acupuncture doit être rayée de la liste des agens propres à provoquer ces évacuations."

believe, that considerable inflammatory action must have given rise to such exquisite nervous sensibility.

Mr. Berlioz in speaking of the diseases to which this remedy is applicable, says,

> "vague and wandering Rheumatism sometimes attacks the external muscles subservient to respiration; the patient is obliged to remain motionless; every motion of the trunk compels him to cry out; a deep inspiration is very difficulty and coughing occasions such cruel pains, that expectoration is impossible. Acupuncturation dissipates instantly this state of distress, and renders to the muscles their full liberty of action. In the space of one or two minutes, a patient whose sufferings drew from him tears, exclaims he is quite cured."

These observations of Mr. Berlioz are fully substantiated by the experience of Dr. Haime of Tours, who has devoted much time and attention to the operation of Acupuncturation, and has lately published a most interesting paper upon the subject in the 13th volume of the "Journal Universel des Sciences Médicales," at Paris.[1]

The doctor declares that his own practice bears evidence of the fidelity of the preceding remarks of Mr. Berlioz. He accuses the Japonese and Chinese, (to whom this operation he says is peculiar of practising,) practising it too extensively, which has been partly the cause of its being disregarded by Europeans, and acknowledges that it was to Mr. Berlioz's cases, which he has related in his "estimable

---

[1] Notice sur l'Acupuncture et observations médicales sur ses effets thérapeutiques.

work" that he owed the fortunate application which he has made of this measure.

The following cases are given by Dr. Haime, which he says support the Theory of Mr. Berlioz.

"Antoinette Boulard, 38 years of age, had experienced in April 1818, a severe attack of Rheumatism, which fixed on the inferior part of the left side of the chest; it gave way in 48 hours to the use of some sedatives, the tepid bath, and the application of a blister to the part in pain.

"Six weeks afterwards I was called to see this woman, who had fallen again into the same state. I found her with the trunk in a state of inability of action, the motion of the respiratory muscles extremely difficult, and the plaintive tone of voice indicated the violence of the pain, which drew from her cries on the least motion. The pulse was small and concentrated, but without sensible acceleration; the body was covered with cold sweats; and the unhappy patient, altogether, was in a state of inexpressible anguish. I thought it right to have recourse to the same remedies which had been successful on former occasions; but my hopes were deceived Three days were passed in this state, and Antoinette obtained no relief: I determined therefore to practice acupunctuation. I introduced a needle[1] at the inferior margin of the cartilages of the false ribs. The instrument had hardly passed to the depth of a few lines, when the patient said the pain

[1] Une Aiguille d'Acier, conique, aigue, longue d'environ trois pouces, et garnie de cire d'Espagne vers son œil, pour tenir lieu de tête.

had changed its seat, and was descended into the abdomen, at the same time that it had lost much of its violence. I continued the introduction to the depth of an inch; by this means the pain was driven from the abdomen, and permitted the patient to breathe freely: however I maintained the needle in its place for five minutes, and then made a second puncture, and successively a third, in the place where the disease had taken refuge. This third puncture made the pain totally disappear, and the patient cried out that I had restored her to life. Sleep of eight hours duration and a state of perfect calmness succeeded this operation.

"However Antoinette sent for me on the following day, saying her sufferings had returned, but with less violence, and entreated me with much earnestness that I would repeat the operation "seeing" she said, "that it was only the sound" (for so she named the needle) " which gave her relief." The operation was this time still more successful. The treatment was now continued for four days, and the last puncture so entirely relieved the pain, that it has not since returned."

In addition to the above successful case the doctor adds another not less so.

"A woman had suffered for several days with wandering Rheumatic pains, which continued daily to encrease in violence; there were however at all times fixed pains in the shoulder and in the right arm, which acquired such a degree of intensity by intervals, that the patient could not refrain from crying out. She was in this state when she came to consult me: finding, however, neither alteration in the pulse, nor encrease of heat, nor redness of the skin, nor tension, nor swelling in the part affected, I considered the case to be simple Rheumatalgia, and passed the needle to the middle of the arm, between the fibres of the Triceps Brachialis muscle; the place designated by the patient as the seat of the pain. The pain was driven into the fore arm, and the second puncture caused it to descend into the hand, and a third being made in this part, caused it totally to disappear, and the patient said with delight and astonishment, she was cured; and was so satisfied with this treatment, that she spoke of it to every body. I have not since seen her, although I requested her (and she promised) to return in the event of a relapse."

But the most remarkable case which has occurred to exemplify the triumphant effects of acupuncturation, was that of a girl of 24 years of age. She was naturally healthy and robust, and had enjoyed good health till she was 15 years old, at which time the signs of puberty were manifested.At this period the system became much disturbed, menstruation was established with difficulty, and continued with irregularity; she lost her cheerfulness, and symptoms of the nervous temperament became predominant. — Various nervous symptoms now evinced themselves, and amongst others an obstinate vomiting: occurred, which subsided only during very short intervals. She continued in this state for two years.[1] From this time she gradually got worse, and in addition to the habitual vomiting which she had endured from the age of 16, she suffered extremely from violent general convulsions. Some medicines were now given which relieved the sickness, and the use of the cold bath suspended the convulsions. After

---

[1] "Ou" says Dr. Haime "la malade contracta l'habitude de l'onanisme et s'y livra sans réserve."

the treatment had been continued two months, she was visited by Dr. Haime, (to whose description of this interesting case, I am indebted for these particulars,) who found her labouring under partial convulsions, with a disposition to vomit occasionally. The means which had been before employed were still continued, but the symptoms became more aggravated, but were a little subdued by blood-letting from the saphena vein. The convulsions were almost wholly removed, at least had become only partial; the spasmodic efforts being concentrated on the diaphragm and stomach; but a nervous hiccup supervened which acquired such a degree of intensity, that the unhappy patient experienced no intervals of ease. All the known antispasmodic remedies were now tried during the space of six months without any benefit. Blisters to the pit of the stomach afforded no sort of relief, and the cold bath gave but a short and temporary alleviation. Scarifications followed by the application of a cupping glass were made on the side of the Dorsal Vertebrae, and the situations corresponding to the pillars of the Diaphragm, which suspended the symptoms but for a few days: relief was only partially obtained by the cautery, and the hiccup returned with its original force accompanied with such a convulsive affection of the stomach, that this organ appeared to act like a pair of bellows, alternately receiving and expelling large quantities of air. At length when the hiccup ceased, it was replaced by partial convulsions or some other symptoms, and vice versa.

Seeing the want of success of all attempts to cure this obstinate disease, and reduced to the necessity of remaining a mere spectator of its dreadful effects, Dr. Haime consulted every book which he conceived might give him some information by which some other curative measure might be suggested; but his researches were totally unsatisfactory, until he met with Dr. Berlioz's observations upon Acupuncturation, when, not less struck with the curious facts which Dr. Berlioz relates, than with the efficacy which it was reported to possess in nervous diseases, he resolved to try it as a sort of forlorn hope, upon his present patient: he accordingly proposed it to her, and readily obtained her consent to its performance. He communicated his intention to Doctor Bretonneau, Physician to the general hospital, who had seen the case with him, and had often spoken of it; and in his presence he performed the operation for the first time. A needle was introduced perpendicularly at the centre of the Epigastrium, and the two physicians soon became convinced of the astonishing promptitude of the remedy; for the instrument had hardly passed to the, depth of a few lines, when the symptoms vanished as it were by enchantment. The operation not appearing to be painful to the patient, the introduction of the needle was continued to half its length, in depth from twelve to fifteen lines, where it was suffered to remain for five minutes. The result was a perfect calmness, and a total suspension of the hiccup for three days, when the same symptoms returning, the needle again was had recourse to, and with the same efficacious effect as at first; and the operation was performed again and again, at longer or shorter intervals, according as the symptoms re-appeared, and always with the same advantages. Dr. Bretonneau became convinced of its efficacy by himself performing it several times. The treatment of the case was thus continued, selecting- the part for the introduction of the needle, according to the situation of the symptoms which each operation was intended to alleviate; and Dr. Haime asserts, it never once failed of success; for the convulsive motions of the head, the instrument was passed into the muscles of the neck; into the masseter muscle, to relieve constant gaping; and into the fore arm when these muscles were affected; and thus, by pursuing the disease as it were, the convulsive disposition was entirely removed, and the patient restored to health.

For the fidelity with which I have reported this case, I refer the reader to Dr. Haime's own record in the 13th volume of the "Journal Universal des Sciences Medicates" and should further evidence of the efficacy of this remedy be necessary from other authority, I have but to mention the experience of Dr. Demours of Paris, who has recently confirmed[1] the report of Messrs. Berlioz and Haime. He dwells with particular force upon its good effects in Ophthalmia, for which he directs five or six needles to be passed between the fibres of the supraspinatus muscle. His method of performing the operation I shall presently notice when describing this part of my subject. The following cases which have occurred in my own practice, I shall now lay before my readers, and I doubt not but I shall make it appear that the beneficial effects of the remedy employed, are sufficiently flattering to deserve the esteem I hold it in, and to justify me in bringing the subject into general notice.

---

[1] See the 66th volume of the "Journal Général de Medecine." [1819, 161--165; 377--383.]

[Cases]

Case 1

George Mc'Laughlan, about SO years of age, a Bricklayer by employment, came to my house in November last, supporting himself by a stick in one hand, and resting the other against the wall, as he proceeded. The body was bent at nearly right angles with the thighs, and his countenance indicated acute suffering. He had been attacked, he said, three days before, with darting excruciating pains in the loins and hips; every motion of the body produced an acute spasmodic pain, resembling an electric shock; and the attempt to raise the body to an upright position was attended by such insupportable agony, as obliged him to continue in this state of flexion rather than encounter it by altering his position. There was no more constitutional disturbance than was to be expected from three days and nights of constant pain; the pulse was a little quickened, and the tongue white, but I attributed this derangement to the irritation set up by the pain and loss of rest. I directed him to place himself across a chair for support during the operation, and I immediately introduced a needle of an inch and a half in length into the lumbar mass on the right side of the spine; in two minutes time I observed that he seemed to rest the weight of his body more on his limbs, and in the next instant, without any enquiry being" made, he observed, that he felt his limbs stronger from the "pain having left his hips." He next plainly indicated that the disease was lessened, by raising his body; from which he only desisted, by being desired to remain at rest, through fear of the needle being broken. The instrument having remained in its place about six minutes, the patient declared he felt no pain, and could, if he were permitted, raise himself upright; it was then withdrawn; the man arose, adjusted his dress, expressed his astonishment and delight at the sudden removal of his disease, and having made the most grateful acknowledgements, left the house with a facility as though he had never been afflicted. The relief was no doubt permanent, as he did not return, which he would most probably have done, had he suffered a relapse.

Case 2

William Morgan, a young man in the employment of a timber merchant, felt a violent pain suddenly attack the loins whilst in the act of lifting a very heavy piece of mahogany. The weight fell from his hands, and he found he was incapable of raising himself. He was immediately cupped and blistered on the part; but two days had passed and he was still labouring under considerable pain, augmented violently by every motion of the body. On the third day the operation of Acupuncturation[1] was performed upon the part of the loins pointed out as the seat of the injury, which, as in the former case, dissipated the pains in five or six minutes, and restored the motions of the back. He returned, however, the next day, with the same symptoms as at first, but in a mitigated degree. A needle was now passed to the depth of an inch on each side of the spine, which, as I expected, terminated the disease in a few minutes, and it was with pleasure that I understood the next morning, that the man had gone to his usual employment.

This case illustrates the observations of the French physicians before cited, as to the efficacy of the remedy in injuries of this description: it is true that in my own practice it is a solitary example; but so decisive was the benefit derived from it, that the case proves a powerful corroboration of both Mr. Berlioz's theory and practice.

---

[1] By a needle of an inch and a half in length.

Case 3

Elizabeth Jacks, a married woman, aged 44 years, was admitted into one of the public hospitals of London, in the year 1817, for an enlarged Bursa situated under the Rectus Femoris muscle. Soon after her admission she was attacked with violent pains in the limbs, which continued to affect her with greater or less violence, till the month of October, 1820, when a severe rheumatic state of the back of the head and of the loins supervened; the one preventing flexion of the neck, the other of the back. Her digestion continued unimpaired, the pulse about its natural standard, without hardness or acceleration. Her nights were passed without sleep, and every motion of the body was performed with pain and reluctance. In this state she applied to me, and I gave her antimonials combined with opium, keeping the bowels open with gentle aperients. Under this treatment, she was in some degree relieved, but as she laboured under the impression that nothing could be done to eradicate the disease, she discontinued it after a short time, but in a few days afterwards (Nov. [4th,] Mr. Carpue was requested to see her; he prescribed ten grains of Dover's powder, to be taken every night at bed time: this dose she took twice without any benefit, The pains had now entirely left the parts they at first occupied, and had fixed on the intercostal muscles above and below the seventh and eighth ribs on each side of the chest; whence, to avoid the insupportable anguish occasioned by the action of these muscles in the process of respiration, this function was (or at least appeared to be) wholly supported by the Diaphragm, the abdominal muscles, and the large external muscles of the neck, chest and back. No other force but that of pressure upon the situation corresponding with the interstices of the ribs gave any uneasiness, but on these parts, the slightest pressure produced intolerable pain: this plainly proved that the disease affected the intercostal muscles alone. Peritonoeal inflammation ensued, and the suffering which this occasioned, banished for the time, all attention to the original disease; but no sooner was this removed, (which was effected by the most active means) than the patient found that she was still the victim of an unrelenting- malady, which had now pursued her upwards of three years. Acupuncturation now recurred to me as a probable mean of relieving her from her sufferings. I accordingly introduced a needle between the sixth and seventh ribs, and another between the seventh and eighth of the right side; in two minutes the patient became sensible of relief, and in two or three minutes more, that side of the chest was emancipated from the disease.

The same operation was now performed on the other side, though the good effect was not equally extensive on this as on the right; yet the patient respired now with so much comparative freedom and ease, that she exclaimed, she should "soon be quite well." — The following day but one, there was a little augmentation of the pain on both sides of the chest, but a single needle introduced into each part, entirely removed it. No return of pain after this time visited the right side, but the left, still continued to be attacked; until at length the third introduction of the needle, dissipated it permanently, and the patient has since remained free from the disorder. The needles in every instance were suffered to remain in the part about five or six minutes.

Case 4

Hannah Howard (a female servant in my house) aged 25 years, became in September last the subject of Rheumatalgia. The shoulders, arms, back and hips, were the parts selected by the disease for its wandering peregrinations. Antimonials, Opium, Guaiacum, Hyosciamus, &c. relieved her occasionally, but at the end of three months, metastasis to the heart suddenly took place. I was called hastily to her at this time; she had fainted, and when recovered from the syncope, complained of violent pain about the region of the heart, which she informed me had troubled her more or less for several hours. Her pulse was hard, and beat somewhat about 106 in a minute; but from its extreme irregularity, it could not be measured with exactness; nor if it might, would it have been found, I believe, to have preserved an uniformity within any two given periods; as both its intermissions and its actions of rapid velocity were produced at uncertain and variable intervals. Copious bleeding, blistering, cupping, with the use of digitalis and colchicum, at length removed the disease; and in three weeks she was able to leave London, to try the effect of country air in restoring her health and strength. She returned to town after a short time, perfectly recovered; and continued so, until an exposure to wet brought on another attack of rheumatalgia; which, after variously shifting its seat for several days, now fixed itself on the left side. The remedies which had formerly been of service, were now taken without relief; and the colchicum[1]

---

[1] I have every reason to believe, the wine impregnated with this plant, is of the most medicinal value, when the infusion has been made with the seeds, rather than the roots, as lately recommended by Dr. Williams, of Ipswich. It is the preparation

(which in most cases of rheumatism will be found, after bleeding, more valuable than any other article of the materia medica) was totally inert. The pain had now acquired such a degree of violence, that the slightest motion of the body gave the most exquisite agony; and so intense was this state of suffering, that the patient could not be urged to speak in a tone loud enough to be conveniently heard, through the fear of exciting an exacerbation of pain, which even such slight motions occasioned. I now had recourse to acupuncturation; having introduced a needle through the integuments covering the interstice of the 8th and 9th ribs, at the part corresponding to the junction, with their cartilaginous epiphises, I continued to press it gently forward, by rolling it freely between my fingers. When it had penetrated to about two thirds, its whole depth (an inch) I enquired if she experienced either pain from the puncture, or relief from the disease; she replied, "she scarcely felt the instrument, but that her rheumatism had suddenly abated of its violence;" and to my surprise, this reply was expressed in her natural tone of voice. She added, "that she could now speak and breathe freely," so that I now found her former taciturnity, which I had attributed to moroseness, was banished. I continued the introduction of the needle, and in a few minutes the disease was dislodged, and fled to the back of the chest, near the angle of the ribs. The motions of the shoulder were now restored to their utmost freedom, and I withdrew the needle, and inserted it into the part which had become the seat of the pain, about two inches from the spinal column. The disease soon began to dissipate itself totally; the patient said she was free from uneasiness, and could make a deep inspiration without pain. The

---

which I have found most beneficial, and upon which I could place the greatest reliance.

instrument having been retained in its place five or six minutes, was withdrawn; the chest had regained its full liberty of action, and the utmost variety of flexion of the body could be used, not the slightest inconvenience ensuing. The next day, however, the pain again visited the anterior part of the chest, and I again had recourse to the needle. The operation was completely successful; for excepting a slight darting pain, which occasionally troubled her for a few days afterwards, no symptoms of the disorder remained, and she continues at this time to attend to the duties of her station in my family.

Whilst occupied with the preceding pages, I received the following communication from my friend, Mr. Jukes; which I subjoin as the strongest corroborative evidence of the efficacy of the practice under our consideration.

"Great Peter-Street, Westminster,

February 27, 1821.

My Dear Sir,

In compliance with your request, I send you an account of the effect of Acupuncturation on our friend Mr. Scott.[1] I received an urgent message on the morning of the [18th inst. from that gentleman, requesting I would visit him instantly. I found him in bed; and, with a countenance expressive of much anguish, he informed me, that for three days he had been suffering severely from pain in the loins, which he attributed to leaving a warm room during one of the late foggy nights. Within the last 12 hours it had acquired such a degree of violence that even respiration was insupportable, except the body were fixed in such positions as permitted the least possible motion. An attempt to resume the erect posture, produced violent spasmodic action of the muscles of the back, which appeared to be communicated by sympathy to those of the abdomen and chest, impeding respiration with a convulsive effort; nor could any motion of the body be made without producing this distressing effect. Neither fever nor general derangement was present; the secreting organs of the body properly performing their function, proved the external

---

[1] Mr. Scott first introduced the operation into England.

locality of the disease. In this state of things, Acupuncturation presented itself to us as likely to afford relief, and it was therefore immediately resorted to.

"I applied an exhausted cupping glass upon the integuments, opposite to the second lumbar Vertebra, and midway between this bone and the edge of the Latissimus Dorsi muscle of the right side, which was the part referred to as the most concentrated spot of the disease. As soon as a needle had penetrated to the depth of an inch, a sensation arose, apparently from the point of the instrument, which the patient described as resembling: that which is produced by the passage of the electric aura, when elicited to a metallic point, diffusing itself at first to some distance around the part, and then extending itself up the side to the Axilla. This sensation continued to be felt for the space of a minute, when a violent pain struck into the right iliac region, immediately above, and corresponding with the line of, the Crista of the Ilium. No pain was now felt in the back, except a dull aching of about two inches in breadth on the right side of the spine, extending from the lower part of the neck to the Sacrum; corresponding with the situation and course of the Longissimus Dorsi Muscle. The pain above the hip now began to subside and in the space of three minutes from its commencement, had ceased altogether.

"The uneasiness along the course of the spine still remaining, a needle was introduced about an inch from one of the upper Dorsal Vertebrae, and another in a corresponding situation to one of the lower Lumbar Vertebrae. The pain in the right side was in a few minutes entirely dissipated, and the patient arose, declaring, that, excepting a slight degree of uneasiness on the posterior part of the chest, near the angles of the inferior ribs of the left side, he was completely relieved from the disease. He, however, requested I would pass a needle in this last situation; on effecting which the pain soon left its last refuge, and the patient dressed himself, and left his house in the most perfect health. I have this day seen him, and he assures me that he has not experienced any return of the affection.

"I should have stated that the sensation, described as resembling somewhat an electrical effect, was experienced from two of the needles only; the first and the last of those which were introduced.

"I send you the history of this case without any comment upon the mysterious nature of this extraordinary operation; yet I am convinced there is something more in it than has been hitherto explained. I have, it is true, some notions (not however fixed) as to its nature; but I would not at present venture to detail them, lest the embers of animal magnetism might be rekindled in the discussion, and the operation from being associated with an exploded theory, sink into undeserved and premature oblivion, from preconceived prejudice.

I am, dear Sir,

Your faithful friend,

EDWARD JUKES."

Conceiving that the foregoing cases will be as satisfactory as a larger number would, I shall not trouble my readers with a more minute detail

I could certainly add many others to the list; but to minds open to conviction and truth, no stronger impression would be made by multiplying examples; whilst the sceptical, would "not be persuaded, though one rose from the dead."

The Operation of Acupuncturation Described.

The first step necessary to the performance of this operation, is the selection of a proper apparatus. It is not requisite, however, that our needles be either of gold or silver, as those of the Japonese are; although it is true that the flexibility of these metals prevents the risque of their breaking; but I have not heard of, or seen, any instance of such an accident with the steel needle, which is the material employed in European practice. It may however be left to the discretion of the surgeon, whether he uses the former or not; it is only of consequence, that the extremity should be finely pointed, and preserved so.

Mr. Berlioz uses a steel needle, three inches in length, which has a head given to it of melted sealing wax. This needle is introduced to such a depth as the operator thinks proper, depending* on the part in which it is used, as well as the nature of the disease which it is intended to remedy. If it be intended to puncture any of the viscera, such a needle will indeed be wanted; but it will be seen by the practice of the French physicians, that though they have sometimes thought it right to penetrate the visceral cavities to the whole depth of this needle, yet it is but seldom that more than one inch of it has been sunk into the part. I have not, in my own practice, ventured to use needles of greater length than one inch, and one inch and a half; and the instrument which I use is an ingenious adaptation of a common sewing needle to an ivory handle, constructed by Mr. Edward Jukes, Surgeon Accoucheur to the Westminster Medical Institution (see plate, fig. 1 and 2.)

Dr. Haime, arid I believe the French surgeons who practice acupuncturation, use this long needle (three inches) and Mr. Demours, who appears to be a man of considerable

81

mechanical genius, has lately invented a new apparatus for this purpose. An exhausting syringe is fitted to the side of a cupping glass, which can be unscrewed and removed after the exhaustion has been effected by a few strokes of the piston, leaving- the glass affixed to the part. From the top of the glass proceeds a hollow staff, in which slides (the tube being- air tight) a handle, armed with a three inch needle, which is inserted to any depth the operator chuses.

The theory which Mr. Demours gives in defence of this instrument is, that the sensibility of the part is so much lessened by the conjestion occasioned by the suction of the pump, that the instrument passes without producing the least pain, whilst at the same time it penetrates deeper, and more readily, through the tumefaction occasioned by the tumescence of the sanguineous capillaries and lymphatics. These advantages, he says, being only obtained by the operators ability of passing the needle whilst the surface of the body remains in the state of tumefaction, he contends they cannot possibly be derived from the simple process of affixing a common glass by the flame of a taper, as the tumor subsides the instant the glass is removed.

I do not think it, however, a matter of any moment, whether a cupping glass be applied or not; it may, certainly, lessen the sensibility of the part, and consequently diminish the pain occasioned by the needle; but this is in general so trifling, that no preparatory steps are required to mitigate it; in fact, it deserves so little the name of pain, that the patient is often unconscious of the needle having penetrated.

The Japonese and Chinese drive in the needle by the stroke of a mallet. This instrument, in use amongst the former, is made of ivory, with holes, sunk on its surface in the same manner as a lady's thimble, which prevent the hammer from sliding off when the stroke is given. Such a method is however objectionable, as well from the danger there would be of breaking a needle not possessing flexibility, as from its being more painful to the patient.

The method to be employed is the following:

The handle of the needle being held between the thumb and fore finger, and its point brought into contact with the skin, it is pressed gently, whilst a rotatory motion is given it by the finger and thumb, which gradually insinuates it into the part, and by continuing this rolling, the needle penetrates to any depth with facility and ease. The operator should now and then stop to ask if the patient be relieved; and the needle should always be allowed to remain five or six minutes before it is withdrawn. This mode of introducing the needle, neither produces pain (or at least very little) to the patient; nor is productive of Hemorrhage, which Dr. Haime says arises from the fibres being separated, rather than divided by the passing of the needle; the former of which (the absence of pain) is a point in its favor, which few surgical operations possess.

It is but rare that I introduce more than one needle at the same time, as a greater number does not appear to be more efficacious than a single one. I, however, depart from this rule (as will be seen from some of the cases) when the pain becomes fugitive from the effects of the instrument; which is a most encouraging symptom. In such circumstances, following the disease by introducing the needles where the pain has removed to, has always proved ultimately successful.

Where also the disease is seated in such several parts, which from their anatomical situation, are known to receive their nerves from distinct or opposite departments of nervous origin) or if the disease pervades more organs or muscles than one, which are but little connected as to their nervous relations; then I regulate the number of needles, accordingly as I suppose the several parts may be more or less connected with each other.

The perforation made by a sharp smooth instrument like a needle, is of such a simple nature, that there is little danger of doing any mischief with one of this kind. Dr. Bretonneau, Physi-to the "*Hospital Général*" of Paris, has made a number of experiments on puppies, the result of which is, that the Cerebrum, the Cerebellum, the Heart, the Lungs, the Stomach, &c. may be penetrated without occasioning the least pain or inconvenience.

In one case, where the heart had been punctured, he afterwards discovered an extravasation of blood into the Pericardium; and Dr. Haime asserts, that his experiments prove the doctrine of Mons. Beclard, respecting the elasticity of the arterial tunics, which may be punctured with impunity. One case of this nature occurred to Dr. Bretonneau, where a jet of blood followed the puncture of an artery. The hoemorrhage was immediately stopped, simply by pressure upon the opening-. Dr. Haime says, that he has often, when performing this operation upon the human subject, thrust the needle to such a depth into the Epigastrium, that the stomach must have been pierced; but that it was productive of no more inconvenience than the same operation upon the more simple parts of the body. I should, however, contrary to such high testimony, hesitate much to puncture an artery, as an aneurism has been known to result from a small puncture made by an awl, which required the division of the vessel for the cure.

I shall here close my subject, not without exciting, perhaps, in the minds of some of my readers, surprise that I have not attempted an hypothesis of the operation. I have by no means made up my mind as to the nature of its action, and rather than venture into speculative reasoning, which may be received as doubtful by some, and visionary by others, I prefer preserving a profound silence. The authors whom I have before referred to, have attempted such an explanation; and should opinions of this kind be considered as deserving attention, the enquirers may find them in the paper upon acupuncturation, in the 13th volume of the "Journal Universel Des Sciences Medicales," published at Paris in 1819.

The needles may be obtained at Mr. Blackwell's Bedford-Court, and Mr. Laundy's, St. Thomas's-

Street, Borough.

FINIS.

3. Memoir on acupuncturation – Morand (Bache, Tr), 1825

Memoir

On

Acupuncturation,

Embracing

A series of cases,

Drawn up

Under the inspection of M. Julius Cloquet

Doctor of medicine.

Translated from the French,

By

Franklin Bache, M.D,

2016 (1825)

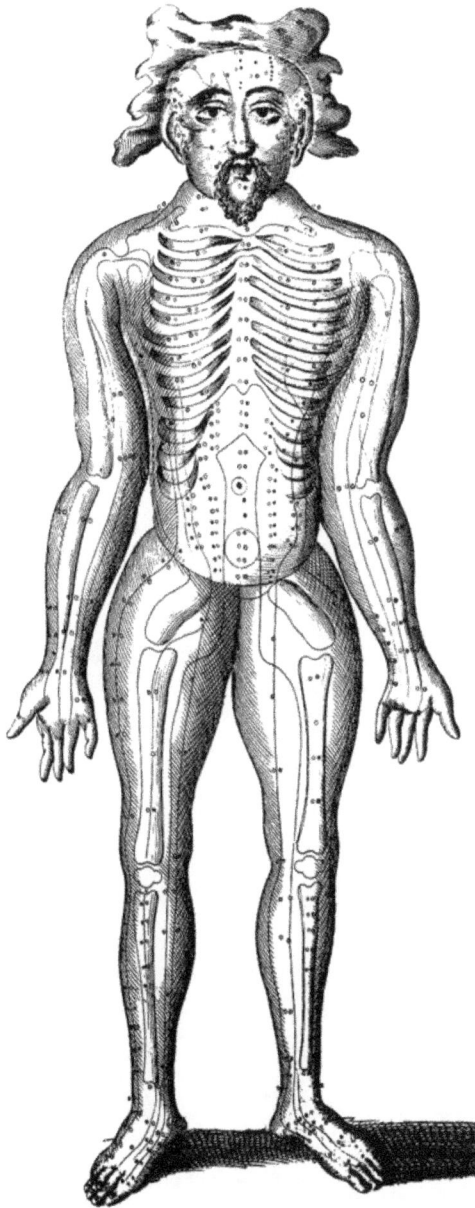

Figure 9 – W. ten Rhyne, Fig. 2, 155.

By the translator

The original of the present Memoir was put into my hands by my friend, Dr. Samuel Brown, late Professor of the Practice of Physic in Transylvania University, with a request that I would prepare a translation of it for the American medical public. I readily acceded to his wishes, believing, with him, that a short treatise on Acupuncturation, from the growing importance of the remedy, and the great attention bestowed upon it in France, could not fail to be an acceptable present to American physicians.

The present Memoir, after an historical introduction, proceeds to give an account of the manner of using the needle, and concludes with a number of cases, illustrative of its effects. Without going too much into detail, it imparts every requisite information, to enable any practitioner to employ the remedy.

Philadelphia, October, 1825.

Introduction

Acupuncturation, a remedy borrowed from the Medicine of the Indians, receives at the present moment general attention.

The academies and departments of Medicine, labour with a zeal and ardour, scarcely credible, to verify the new results of M. Julius Cloquet. At the same time, the most distinguished philosophers are multiplying researches and experiments, to explore the new road which has been opened to them.

The journals, appropriated to the publication of clinical facts, observed in the several hospitals, fill their columns with observations and reflections on this new method of curing diseases.

Even men of the world are not content to remain ignorant of this triumph of Medicine in our days: in short, every where acupuncturation is spoken of; in the drawing room, as well as at the Institute. The enthusiastic physician hopes, by means of a needle, to combat diseases the most inveterate and rebellious. It will become, according to his views, *supremum contra omnia mala, remedium.*

Another, ingenious in building theories and creating systems, is persuaded, that he finds, in the phenomena exhibited, a satisfactory solution of questions the most obscure; questions, which have been agitated for centuries, and to which the attention of the greatest men have been directed for so long a time, and to so little purpose.

Others again, by whom incredulity is erected into a system, refusing to believe what is demonstrated in the most satisfactory manner, and filled with a proud disdain, resign, without previous examination, the practice of acupuncturation to the people with whom it originated.

But, happily, there are wise and enlightened men, true friends of science, who, silencing the voice of the passions, tranquilly meditate the observed facts. Disdaining the pomp of theories, and constantly fearing to stray from the right path, by entering the vast domain of hypothesis, they follow, with perseverance, the straight road of observation, note facts with scrupulous exactitude, and afterwards report them with a candour and good faith, but too rare in our days. It is to such men, that truly belongs the name of philosophers; it is to them that science and the arts owe all their progress. They are models for the imitation of young physicians, and the guides which I have chosen. Like them and with them, I wish to say:

Amicus Plato, magis arnica veritas.

A pupil of M. Julius Cloquet, to whom the honour belongs of having revived acupuncturation amongst us, and of having made so happy an application of it in the treatment of diseases, I resolved to select this therapeutic agent, as the subject of my inaugural dissertation; to submit to my judges my opinions, and the observations, I have been enabled to make, as well at the hospital of St. Louis, and the Hôtel-Dieu, as at other places, where I have seen it put in practice in a great number of instances.

At the same time, while giving this simple exposition of facts, I am happy to have an opportunity of rendering homage to the merit of the young and learned preceptor, under whose auspices I enter upon the career which I pursue, and whose kindness and precious friendship I have constantly continued to enjoy.

Division

I shall divide my subject into three parts:

In the first part, I propose to make known what had been observed on acupuncturation, before the labours of M. Cloquet, comprising the experiments, observations, and opinions of the various authors, who have treated of this subject.

In the second, I shall give a description of the operation, noting with care the phenomena that present, and the theories to which they have given rise.

In the third part, comprising a series of cases, I shall speak of the diseases, against which acupuncturation has been employed, the parts on which it has been tried; and, finally, the results which its employment has furnished

Part I

Definition

Acupuncturation, *acupunctura*, derives its etymology from the Latin, *acus*, a needle, and *punctura*, a puncture. The operation consists in causing a needle, (without regard to the metal of which it is made,) to penetrate into some part of the hody, cither of man or animals.[1]

If we seek to know, in a positive manner, the origin of acupuncturation, it will be found lost in the obscurity of remote ages. But if it be impossible to determine, at what epoch it was practised for the first time, we are, nevertheless, certain, that, unknown to the Greeks, Romans, and Arabians, it was employed from time immemorial in China, the country which we regard as the cradle of the world; and that the people of that country transmitted it to the inhabitants of the island of Coree from Japan. It is, says Vicq-d'Azyr,[2] according to *Ten Rhyne*,[3] very much employed in the last mentioned country.

It is to the latter author, and to *Kœmpfer*,[4] that we owe the greater part of what we know on this subject.

---

[1] [See also Elliotson, Acupuncture, in *The Cyclopædia of Practical Medicine*, 1832, 1:32-34.]

[2] [See Vicq-D'Azyr, *Encyclopédie Méthodique*, 1792 ; Diderot, *Encyclopédie Méthodique*, 1830, 152:450.]

[3] [Coroll. De Acupuncturâ. Hist. de la Chirurgie ; Dissertatio de arthritide, mantissa schematica, de acupunctura et orationes tres.]

[4] [Engelbertus Kempfer, *History of Japan*, Scheuchzer (Tr.), London, 1727.]

Accordingly, it appears, that it is only within a century and a half, that the operation has been known in Europe.

But let us see, in what manner, and under what circumstances, the Chinese practise the operation.

The Indians, says *Kœmpfer*, employ two instruments in performing acupuncturation, namely, a needle and hammer. They give the name of *tentassi*, or *exploratores*, to the men, who know the places where it is proper to place the needles. As to those who practise the operation, they bear the name of *foritatte, jussa explorantis faciunt*. This means is so much employed, and they have so great a veneration for acupuncturation, that this class of physicians never leave their houses, without carrying their instruments. The hammer they employ, according to the author of the article Acupuncturation, in the Dictionary of Medical Sciences, is made of ivory; while, according to Kœmpfer, it consists of horn. One of its extremities is larger, rounded, and pierced with a number of small holes, similar to those which cover the surface of a thimble; it contains lead, enclosed in its interior, in order to add to its weight. The handle is long and hollow, for the purpose of containing the needles.

The needles are of two sorts. The first kind are about four inches long, and very slender, especially towards the extremity intended to penetrate the tissues; while the other extremity is fastened into a handle, three inches long, rounded, shaped in spiral, in a word, like the thread of an elongated screw. The second kind are without handles: their length is the same as that of the first kind, and thickness equal to that of a harp string. No other metals hut gold and silver are employed in the fabrication of these instruments, and no one in China is allowed to make them, without the special authorization of the sovereign.

With these people, the operation of acupuncturation is practised according to two different methods. In the first, the needle should be seized with the left hand, between the thumb and fore finger, supported by the middle finger, and then placed on the part intended to be pierced. The hammer being taken in the right hand, they give the needle one or two blows, to make it traverse the cuticle and true skin; after which, turning it rapidly between the fingers, they sink it perpendicularly to the depth of from half an inch to an inch, according to the part on which the operation is performed.

In the other method, the needle only is used, it being made to penetrate by the second manipulation above mentioned; in other words, by rotating it between the thumb and fore finger. The spot chosen to place their needles, is that, in which the pain is most acutely felt. They often multiply the acupunctures, inserting at the same time several needles, always observing to leave an interval of an inch at least between them.

According to Kœmpfer, the time they are allowed to remain is very short; being equal only to the duration of an inspiration and expiration. The operation should be repeated, until the pain has completely subsided, being always careful to avoid lesions to the nerves, arteries, and tendons.

It is especially in combatting a very painful colic, called senki, that they have recourse to the operation. The description of this disease is too curious not to be given in an abridged form. It is endemic, and so common, that it is a rare thing to meet an individual, somewhat advanced in years, who has not suffered an attack of it. Strangers are not less subject to it than natives. The causes are stated to be the climate, food and drinks. The principal symptoms are very sharp abdominal pains, with swelling of the belly, spasms, and tenesmus. The patient is also affected with suffocations and convulsions. The sensation is such as might be supposed to be produced by the tearing of the tissues from the groin to the ribs. The pain experienced is analogous to that, which would be produced by repeated thrusts of a poniard. Such is nevertheless the train of symptoms, which are made to disappear instantaneously by several needles, placed in the following manner: — One above the umbilicus, a second on a level with it, and a third below it. Others are often placed on the sides of the abdomen.

But how serious are the accidents, if recourse is not had to this remedy! The disease, after having produced long sufferings, gives rise to tumours upon the surface of the body, swellings of the testicles, suppurating fistulas, abscesses in the anus; and in women, suppurating tubercles, either in the vulva or groin.

If we consult Ten Rhyne, in how many other cases do we find acupuncturation praised. According to this author, the needle should not always be confined to traversing the skin, but made to reach the intestines, and the womb even in pregnant women. In no case, says he, do accidents occur. It is employed, with success, in all kinds of colic; in diseases of the head, and headache, both recent and inveterate; in soporose diseases, epilepsy and ophthalmia; in diarrhoea, cholera morbus, dysentery, and anorexia; in flatulent affections, hysterical paroxysms, lippitudo, incipient cataract, and coryza; in continued fevers.

intermittents, verminose affections and tetanus; in fine, in all convulsive diseases. These details, collected by Ten

Rhyne, are to be found narrated by Vicq-d'Azyr. (Treatise on the Physiological Sciences, article on Acupuncturation.)

This last author makes four classes of tire diseases, against which acupuncturation is proposed to be employed:

    1st. Soporose affections; *Comata*.
    2d. Convulsive affections; *Spasmi*.
    3d. Pains; *Dolores*.
    4th. Diseases attended by discharges; *Fluxus*.

In the same work, a description is given of a very painful colic, relieved by a single acupuncture. Vicq-d'Azyr has drawn this case from Ten Rhyne, and relates it as follows: A guard of the Emperor, after having performed a long march exposed to the sun, being tormented with an ardent thirst, drank with avidity, a very irritating species of beer, similar to the *Schnops* of the Germans. A short time after, he felt over the whole abdominal region, very sharp pains, together with tenesmus, and contractions. Acupuncturation being performed, he was not merely relieved, but cured immediately; and the pain never afterwards returned. Vicq-d'Azyr, who, as I have already mentioned, has written a memoir on this subject, docs not state that he ever practised the operation. He proposes to class it among irritants or stimulants. According to him, its action may be compared to that of moxa, blisters, and cups. They are deceived, says he, who assert, that a needle may traverse the womb with impunity; because, in operating, it is very difficult to determine to what point it may have reached, or to ascertain, with any degree of certainty, what organs may be involved.

Dujardin, in the first volume of his History of Surgery, gives a repetition of what may be found in *Ten Rhyne*, from whom nearly all the authors on this subject have borrowed.

I have thought it proper, however, to note from this writer the following propositions:

1st. Acupuncturation succeeds perfectly in cases of pregnancy, when the foetus, moving continually in the womb, (without regard to the cause of its movements,) occasions sharp pains to the mother. It is necessary, under these circumstances, in introducing the needle, to cause it to traverse both the womb and foetus. By proceeding in this way, all possibility of accident is guarded against. The author does not state, whether it is through the vagina, or by piercing the walls of the abdomen, that the operation is to be performed.

2d. In feeble persons, it is most proper to practise the operation on the anterior part of the body, and on its posterior part in strong persons. This direction is evidently very vaguely expressed, and may be considered as unimportant.

3d. If the patient's pulse cannot be felt, the needle ought to be introduced in the vicinity of the veins. By this mode of proceeding, the arteries are liable to be wounded. It consequently appears, that the advice given to avoid these vessels is disregarded.

4th. In adults, the needle should be inserted to a greater depth than in children, or old persons; and in persons enjoying a degree of plumpness, than in thin persons.

5th. Finally, the needle ought to remain in the tissues, during the interval of time, occupied in the performance of thirty inspirations, and as many expirations. In the event of the patient's inability to support its presence for this length of time, the operation is to be repeated five or six times.

Upon reading Berlioz, (Memoir on Acupuncturation, published in 1816,) we find, not merely theories, and observations obscured by the night of time, but many practical facts, and verified and authentic results.

The success, which lie announces, is so great, that it is impossible to explain the reasons, which have prevented a repetition of his trials. According to M. *Berlioz*, a needle, inserted to the depth of several lines, either removes immediately, or diminishes pain; and it is extremely rare for it to resist three or four operations.

I presume it may not be without its use to report the observations found in this work. The subject of the first is a young girl, attacked with a nervous fever, the consequence of a violent and prolonged fright. The disease had existed for two years. Treated by cinchona, the symptoms, far from being mitigated, were increased. A multitude of remedial agents were employed, but without success. The young patient was observed to waste sensibly. Not knowing how to oppose the disease, acupuncturation was proposed, and practised immediately. A very long sewing needle was employed, covered near its eye with wax. The young patient (for it was she herself that performed the operation) introduced the needle, at first perpendicularly, and then obliquely, through the parietes of the abdomen, over the epigastric region. From the moment the first puncture was made, the pains ceased, as if by enchantment, and the calm was complete. The paroxysm did not reappear this, or the succeeding day. On the second day after the operation, slight traces only of the disease remained. To prevent the return of the fever, a new acupuncture was performed on the third day, and repeated once in three days, for the space of two months. At the end of this time, the remedy appeared to lose its effects hy habit, and was resorted to oftener, namely, twice a day for six months. From this time, the disease was cured. But three months afterwards, there occurred a new attack of fever, when acupuncturation was again resorted to, and a new cure effected, which was now permanent. Dr. Berlioz states, that he employed, at the same time, preparations, containing opium. Accordingly, it may be asked, whether it was to the acupuncturation, that he owed the success, which he obtained.

The facts, which he afterwards details, appear to me far more conclusive. A peasant, aged 40 years, had been afflicted for two months with a convulsive cough, attended with pain in the epigastrium. A milk diet, and the exhibition of opium had relieved him considerably. Nevertheless, walking, and especially exercise on horseback, produced much fatigue, excited cough, and gave rise to pains in the region above mentioned. A single needle, introduced into the epigastric region, and inserted sufficiently deep to reach the stomach, caused a sudden disappearance of every symptom, and, what is more important, the cure was permanent. In this case, the needle remained hut three minutes.

The case, which follows, is that of a man, who fell backwards from a height of 10 or 12 feet, upon stones, causing numerous contusions over the posterior part of the trunk. Placed on a bed, it was absolutely impossible for him to perform any movement. Eleven punctures, performed on the back part of the neck, enabled him to raise his head. The same operation, being repeated on the following days upon all the bruised parts, cured them very rapidly. The author of these observations recommends to have recourse to the needle, in every species of pain, whether rheumatic or nervous, and in intermittent fevers, at the moment of the paroxysm. He does not fear the lesion of the most important organs by the puncture. According to him, apprehension is injurious to the effect of the operation, and the number of needles is unimportant, not increasing the efficacy of the remedy; hence a single one is ordinarily sufficient.

Dr. *Berlioz*, however, was not content to report his success with this operation, but wished to ascend to its cause. He felt desirous of explaining the mode of action of the needle, and he does so in the following manner: — It is not by substituting-, for one irritation, another which is stronger, as has been advanced, that acupuncturation acts. For in no case is the success so marked, as when the introduction of the needle gives little or no pain. The remedy is a stimulant to the nerves, to which it restores a principle, of which they were deprived as an effect of pain. He concludes by proposing to charge needles with the electric fluid, with the view of rendering their action more powerful. He also employed two needles, made of different metals, connected together by means of a third. Their action did not give more favourable results.

These observations are doubtless very curious; but the following are not less so, drawn from the practice of M. *Bretonneau*, chief surgeon of the hospital of Tours, a practitioner, whose talents do the greatest honour to modern surgery. The facts, communicated to the society of Medicine by his correspondent Dr. *Haime*, have been given to me in a still more exact manner by M. *Velpeau*, formerly a pupil of M. *Bretonneau*, whose practice he follows with the zeal and ardour, which distinguish him at the present day.

A young female, twenty-one years of age, had been tormented for a long time with a continual hiccough. The affection having been treated by antispasmodics and revulsives, for many months, and always without advantage, the surgeon of Tours, prompted by an inspiration altogether medical, thought of having recourse to the means so much cried up in China and Japan. Like *Berlioz*, he employed instruments, different from those in use amongst those nations. He took a steel needle, about 7 or 8 inches long, slender and very sharp, and introduced it, through the parietes of the abdomen, into the epigastric region, so as to reach the stomach. The pain produced by the introduction was far from being acute. The needle had hardly traversed the tissues, when the hiccough ceased. After remaining several minutes, it was withdrawn. No accident occurred, either during or after the operation. During the whole of this day, the disease did not recur; and it was not until twenty-four hours afterwards, that she felt it again. From that time, a new acupuncture was decided upon, and practised at the same point. This time, the needle was made to penetrate to a still greater depth, even to the vertebral column. The hiccough was suspended for the space of forty-eight hours. After these first successes, care was taken to persevere in the remedy, and very shortly, acupuncturation, at first only palliative, effected a complete cure. This fact is curious, not merely from the termination of the disease, but particularly because the needle had traversed, without the smallest accident, organs, whose lesion is considered mortal. Some doubts may be entertained as to the cause of this cure, when it is known, that the patient was influenced at the same time by moral causes; and that by just remonstrances, a termination was put to a habit, which she had contracted a long time before, (a circumstance not noticed at first) of subjecting herself to practices, equally inconsistent with morality and health. But the cessation of the hiccough was spontaneous; occurring as soon as the acupuncturation was practised. It was, however, only temporary, and the hiccough was only removed by new operations; while the discontinuance of the habit of the patient, could not produce advantageous effects, until after the lapse of a considerable time. So that, all that can be admitted is, that this circumstance hastened the progress of the cure, and prevented relapses.

Having penetrated the stomach, that organ, to which physiologists have attributed so exclusive an agency both in health and disease, M. Bretonneau wished to ascertain how serious the practice of acupuncturation might be to other organs, not less essential: — I mean the heart, the lungs, and the brain. He accordingly made experiments, which gave the following results.

Having taken six young dogs of a vigorous breed, H introduced a needle into the substance of the brain itself. Causing it to penetrate at first by the anterior fontanelle, he directed it from before backward, as far as the occiput. Having withdrawn it almost immediately, he buried it under the inferior and anterior angle of one of the parietal bones, giving it a transverse direction to the opposite side, and then withdrew it, as in the first case. Three of the animals gave evidence by their cries, that they did not remain insensible to this new stimulus. The three others put forth no cry, and did not even seem to perceive it. None of the animals experienced any accident, as a consequence of the operation. The experiment was repeated many times, without producing any inconvenience.

The same needle was afterwards introduced into the animals, through the parietes of the chest, over the region of the heart, and sufficiently deep to involve this organ. That the experimenter attained the object proposed, was evinced by the circumstance, that, at each contraction of the heart, the needle underwent movements of elevation and depression, in correspondence with its systole and diastole. Three supported this new acupuncture without accident. A fourth sunk under its effects in half an hour. On opening the body, a little blood was found extravasated into the cavity of the pericardium. The fifth became drooping, and weak, and appeared unable to survive. Having killed him, blood was found, as in the preceding case, in the pericardium, but in less quantity. In the sixth, the needle did not reach the heart, but the lungs were penetrated: the animal sustained no injury. There is every reason to believe, that, if very fine needles had been employed, similar to those made use of by M. *Cloquet*, none of these animals would have fallen victims to the experiment. After having penetrated the heart, the same practitioner pierced arteries of all calibres, without serious consequences in any one instance. Finally, his confidence became so great, that he did not hesitate, on his own person, to introduce the point of a very fine needle into the sides of the brachial and radial arteries.

M. Bretonneau had occasion also to resort to acupuncturation in a case of rheumatism of long standing, which disappeared in summer, but returned again, whenever the weather became cold or moist. This rheumatism had its seat in the deltoid, and extended over the whole of the posterior part of the shoulder, causing much suffering to the individual affected. Many remedies having been fruitlessly used, he employed the needle. It was inserted over the point where the pain was felt with the greatest intensity. The first introduction caused the pain immediately to cease; but it recurred twenty-four hours afterwards, though much less acutely. Acupuncturation was again resorted to, and practised as often as the pain returned. Finally, when the remedy had been used seven times, the pain disappeared entirely, and the cure was complete. Whether it subsequently returned, I am unable to affirm.

These successful trials seemed to promise a frequent employment of the needle. The observations made were collected and reported, since when they have remained in obscurity. The sick profited by the benefits of the operation, and, as it too often happens in society, the benefactor was forgotten.

Two years ago, M. *Velpeau* repeated some of these experiments. He passed a very fine needle through the walls of the heart of a dog, and allowed it to remain for several minutes. The animal did not give any sign of pain. The needle being withdrawn, not the least accident occurred, and the dog lost nothing of his vigour. He was kept confined for several days, after which he made his escape.[1]

I have seen a cat, whose chest was pierced entirely through, from one side to the other. The animal remains to this day full of life.

The author of the article *Acupuncturation*, in the new Dictionary of Medicine, M. *Béclard*, is far from recommending the practice. He says, that, before studying the therapeutic agency of this singular operation, it is proper to ascertain with exactness, the effects of the puncture upon different parts of the body.[2]

M. *Béclard* adds, that if the deepest punctures, even such as involve the viscera, do not always produce accidents; yet

---

[1] [Velpeau, *Traité d'anatomie chirurgicale*, 1:544.]
[2] [Adelon Béclard, Dictionnaire de Médecine. 1 :335-336.]

they give rise sometimes to very serious ones, and even to death itself.

The puncture by rotation, according to him, is less painful, and followed by fewer accidents.

Finally, he concludes with this expression. "Before having made any experiments with this operation, and before it came into use as a curative agent, I was sufficiently disposed to believe, that it would be best to leave it to its inventors. Experience has confirmed me in this opinion."

I acknowledge that I am far from sharing the opinion of this learned professor; and the facts, which I have recorded in the third part of this dissertation, are far from being in unison with his assertions.

Part II

GUIDED, no doubt, by a knowledge of most of the facts which I have detailed, M. Julius Cloquet[1] determined to revive acupuncturation, to verify the assertions of the authors, who had employed it, and to demonstrate, by a series of experiments, its advantages and inconveniences, — in a word, to ascertain the rank, which the operation should occupy among therapeutic agents; well satisfied that it could not be attended with much inconvenience, even though it might not possess all the virtues which have been attributed to it.

It was in the treatment of rheumatic pains, that recourse was had to the operation in the first instance. Having succeeded, even beyond his hopes, he multiplied his researches, which were rewarded with new successes. Soon afterwards, he used it in other diseases.

Being no less fortunate in these trials, and being placed on a proper theatre for extensive observation, (*the wards of Surgery and Consultation of the Hospital of St. Louis*), he practised acupuncturation in innumerable instances. But it was not enough for him to have his labours crowned with success 5 he desired to mount to the cause. Shortly afterwards, he thought he observed phenomena, which had escaped the notice of his predecessors. After satisfying himself, that he was not deceived, he prepared a memoir

---

[1] [See also Vannes, *Traité de l'Acupuncture* [Treatise on Acupuncture, composed from the Observations of M. Jules Cloquet, and published under his inspection], 1826; and Morand, *Memoir on acupuncturation, embracing a series of cases, drawn up under the inspection of M. Julius Cloquet*, 1825.]

on the subject, which was presented first to the Royal Academy of Medicine, and afterwards to the Institute.

The following are the fundamental propositions contained in this memoir.

1st. Acupuncturation acts decidedly on pains, without regard to their cause, or seat.

2d. Some pains disappear without returning, while others reappear after a variable time. In the latter case, they are always weaker than before the operation, and may be removed again, by one or more new acupunctures.

3d. Some pains are diminished only, without being made to disappear entirely; while others are translated.

After demonstrating, that, in acupuncturation, the needle is charged with electricity, he concludes by asking, whether pain in any part does not depend on an accumulation of the electric fluid in such part; whether, in short, the proximate cause of all inflammation does not consist in the accumulation of the same fluid; and whether, by means of conducting wires, connected with the needles, we should not augment the power of the remedy.

Having followed most of the experiments of M. Julius Cloquet, having seen acupuncturation practised a great number of times, and performed it myself, I propose to detail the facts, which have struck me as the most important, preceding this exposition by a description of the operation itself, and of the phenomena to which it gives rise.

In performing acupuncturation, M. Cloquet at first used a common sewing needle, the superior extremity of which, or that intended to receive the thread, was surrounded with sealing-wax, forming a small oval handle, to facilitate its introduction. Soon afterwards, he employed a needle, furnished with a little ivory handle, turned in the form of a lengthened screw, precisely similar to that described by Kœmpfer, and which is used in China, except that the needle was of steel. A short time afterwards, this instrument was modified by Professor Recamier. This professor caused needles to be made of different lengths, after which he mounted them on a species of port-stone, (porte-pierre.) The base of the needle is placed in a cavity, and retained there, by being pressed with a screw, which may be tightened at pleasure. Pressing on the head of the screw, this is made to mount or descend in a groove, formed in one of the sides of the port-needle, by the same mechanism, by which the blade of a knife is drawn into, or made to protrude from, the handle. The needle being retained by the pressure of the screw, it may be made to protrude, or can be concealed, either entirely or in part, from the patient, on whom we e are about to operate. When we wish to perform acupuncturation, we place the port-needle over the point intended to be punctured: then pressing upon the screw, the point of the instrument is made to penetrate the tissues, perpendicularly or obliquely, either with or without rotation. When it is judged that the needle has penetrated sufficiently deep, the screw is turned in the opposite direction. The base of the needle being no longer pressed, the port-needle may be withdrawn, to be subsequently employed in the same way. This method, certainly very ingenious, has the advantage of not intimidating the patient, and is suitable, when we propose to operate on timid persons; but it has the disadvantage of rendering the operation complicated, and at the same time longer, and therefore more painful.

The same practitioner has employed needles of gold; but as it is impassible to render them as slender and sharp as those of steel, without becoming too flexible, I incline to the opinion, that it is best to reject them, and make use of those which we shall now describe.

Having remarked phenomena, which led him to suppose the accumulation of the electric, or some other fluid, upon the needles, during their insertion, M. Cloquet gave to the instrument the following form. His needle is composed entirely of steel, its length being different, according to the depth, it is proposed to penetrate. The degree of slenderness to be given to it, should be regulated by the importance of the organs through which it is to pass. One of its extremities, to serve as a handle, has the form of a very small olive, being rounded for the length of from three to eight lines, according to the length of the needle. Its diameter is about a line; and it is surmounted with a small metallic ring, intended to receive a conducting wire for the electrical or nervous fluid. The body and point arc made as fine as possible.

The spot, where the instrument should be placed, is that in which the pain is most severely felt; but in case the nature of the tissues does not permit this, then the introduction should be effected at the nearest safe point. This general rule has but one exception, and that is furnished by the neuralgies, in which the neighbouring part is most proper for the operation. In the practice of acupuncturation, it is proper to avoid the large nerves, as well as the joints, and although the experiments of M. *Bretonneau* may be very conclusive, it is prudent to avoid the perforation of arteries.

The part, into which the needle is to be inserted, being well determined, the skin is to be rendered tense; for if this precaution be omitted, especially in subjects, whose tissues are loose and flaccid, the integuments, becoming twisted round the needle, will render its introduction more difficult and painful.

The needle being held in the right hand, between the thumb and fore finger, is inserted perpendicularly, and afterwards obliquely, according to the indications presented by the disease, and the nature of the tissues; at the same time executing a movement of demi-rotation. It is important not to penetrate beyond the part which is the seat of the pain; for observation has demonstrated to the surgeon of St. Louis, that the operation will be attended with less beneficial results. It is especially in muscular rheumatisms, that it is important to attend to this rule; and rather than to pass it, it is incomparably better that the painful muscle should not be reached; for then the remedy would have the same efficacy, or very nearly so. If the directions I have given, are followed, the insertion of the needle will scarcely give pain. Being once inserted, a series of phenomena will be very soon observed, for a knowledge of which we are indebted to M. Cloquet. The pain ceases to be felt, though this does not happen constantly. In some instances, the original pain, or to speak more properly, the morbid pain, ceases also.

This event, however, is not the most usual; but it is only after the lapse of a longer or shorter time, that the pain is observed to diminish or cease entirely. If we approach the finger to the part of the needle, which is external, and touch it slightly, nothing will be perceived. The skin is sometimes elevated round the instrument, preserving its natural colour; soon afterwards it sinks, and a red areola, of an erysipelatous nature, is found to form. The time, at which this last is produced, is far from being uniform. Some* times it occurs in five or six minutes after the introduction of the needle; sometimes not until after an interval of two or three hours; and, again, in other cases, it does not make its appearance at all. The patient next feels pricking sensations, referred to the point of the needle, muscular contractions, numbness following the course of the large nerves, and agitations in the small fibres. All these phenomena are far from being constant; neither do they occur at the same time.

It is not unusual, at this moment, to observe perspirations, covering the part of the skin, corresponding with the organ, which is the seat of pain. Sometimes, though more rarely, the sweats are general. The pain ceases from that time, or is diminished; or otherwise translated, or carried back, diminished in intensity, to some place, more or less distant from the needle. It is also about this time, that syncopes sometimes occur, more or less completely formed, and of variable duration; and this indeed is the only accident, I have observed to happen in the employment of this curative agent. The occurrence of these syncopes, it appears to me, cannot be caused by the pain, produced by the puncture, as it does not happen, until all painful sensation has ceased. I therefore suppose, that there must exist some other cause. Struck with a reflection of M. *Récamier*, made when employing the aspersion of cold water in an ardent fever, that he practised a bleeding of caloric and electricity, I ask, whether the needle does not act as a true lightning rod, introduced into the system; whether, in a word, it does not charge itself with the electric fluid. I communicated this view to M. Cloquet, and he has adopted it; but still it remains to demonstrate its justness, and support it by facts; for, until then, it is merely an hypothesis; and there is much difference between mere probability and truth.

The truth of this hypothesis was proved by M. Cloquet in the following manner. Having inserted a needle into the thigh of a patient to the depth of an inch, about six minutes afterwards, at the time of the formation of the areola, he touched the body of the instrument. Having previously wet the end of his finger with saliva, he perceived, after a few minutes, a slight shock, similar to that produced by the conducting wire of a very weak voltaic pile. At every new touch, the patient complained of severe prickings, and darting pains, proceeding from the point of the needle. If the touch was repeated too often, the sensation was no longer distinct, doubtless because the needle had not time to charge itself anew. By waiting a few minutes, the same effects are reproduced. I repeated this experiment, and was soon convinced that I was not the dupe of my imagination; for, in some instances, the shock was sufficiently strong, to produce a slight numbness in the finger employed.

Professor Recamier 9 to whom I mentioned these results, did not appear surprised, and determined immediately to experiment. He placed two needles in the lumbar region of a patient, at the distance of about two inches apart, and to the depth of about fifteen lines. The individual was affected with muscular rheumatism. Six minutes after their introduction, having touched the needle first inserted, the shock appeared to him quite distinct, and was equally so to several students present. Others, to whom the sensation was not equally evident, either from prepossession, or from their nervous sensibility being less, without denying the fact, refused to admit it. It was proposed to place a small body on the feather of a quill and present it to the needle. A very small piece of paper was employed, which, when at the distance of half a line, flew to it, and afterwards turned round; at least it was supposed to do so. The trial was then repeated, and with the same result.

But in the progress of time, the presence of a fluid became demonstrated by much more conclusive experiments. An electrometer was employed in the presence of Professor Pelletan. The instrument was defective, and the presence of any fluid was not manifested; and had it not been for the perseverance of M. Cloquet, the existence of a fluid, which is now so well demonstrated, would have continued to be denied. The trial was, therefore, repeated, and, employing a very sensible electrometer, the existence of an electric fluid was placed beyond doubt. But is it certainly the electric fluid, or is it not rather the nervous fluid, as it is denominated by M. Cloquet? Finally, if it be the electric fluid, is it positive or negative? These are questions, which have not yet been answered, MM. Pelletan, Pouillet, Edwards, and many other physicians are now occupied with this subject, and, perhaps, will shortly instruct us on these points. M. Béclard, who has made some experiments, says, that, if we insert a very fine metallic wire into any part of the body, provided the other extremity is brought in Contact with a moist substance, galvanic phenomena will be produced. For example, if you plunge a metallic wire into the upper part of the thigh, and the other extremity be brought to the mouth, a moist surface, electricity will be developed, and the current will pass from below upwards.

But if we form an electrical circle, — for example, if we insert one extremity of the needle into the upper part of the thigh, while we make the other extremity penetrate its lower part, — then it becomes completed on one side by the arc of the wire, and, on the other, by the thigh. Electricity becomes disengaged, but we are ignorant in what direction the current takes place.

M. Pouillet, who, as we have already mentioned, is occupied on this subject, after having recognised the presence of electricity, inquires whether the development of the electrical fluid is not due to some other cause than the contact of the metal with the living tissues. According to the observations made by M. Cloquet on the oxidizement of the needles, a chemical effect verified by this learned natural philosopher, this circumstance is an essential condition to the production of electrical phenomena. To satisfy himself on this point, he employed needles, which oxidize very difficultly, or not at all (needles of platinum). He inserted them in the tissues, and not obtaining a disengagement of the electric fluid, was finally confirmed in his opinion.

The same natural philosopher, supposing that the nerves might be considered as electrical conductors, and that their action on the living system was due to electricity, metallic bodies being the best conductors, wished to interrupt the current. For this purpose, he plunged a brass wire into the spinal marrow of an animal, in the cervical region, and the other extremity into the lumbar region; but no electrical phenomena were obtained.

M. Cloquet has remarked, as the result of experiment, that the action of the needle would be more prominent, if connected by its base, with one extremity of a conductor, the other extremity being plunged into liquid muriatic acid, or a saturated solution of muriate of soda. Indeed, when he has proceeded in this way, I have observed the action of the needle to be more prompt, the pain experienced towards its point being more pungent. Sometimes it has been necessary to withdraw for a moment the conductor, and even the needle, to calm the shooting pains, felt by the patient.

This series of experiments naturally leads to this reflection: does the disappearance or diminution of the pain, which is observed, depend upon the withdrawal of the electric or nervous fluid, or does the cause remain unknown?

If I adopt the opinion of M. Cloquet, the question will be answered in the affirmative; for, according to him, there is neither disappearance, nor diminution of pain, without a subtraction of fluid; and it is in conformity with this explanation, that he has laid down this proposition, (under a form, it is true, expressive of doubt), that pain, as well as inflammation, have for their cause, an accumulation of the electric or nervous fluid, in the organs in which they have their seat.

It will be perceived that this proposition has a very extensive bearing.

But Professor Pelletan affirms, according to his experiments, that the action of the needle on the economy is entirely independent of electrical phenomena, or of oxidizement; for, upon employing instruments of platinum or gold, equally beneficial results were obtained from acupuncturation. The practice of the Chinese comes in support of this opinion; for nowhere is acupuncturation more employed, or more highly thought of than in China, and yet all their instruments are made of gold or silver. M. Cloquet has himself made use of needles of silver, and I have not observed that his practice was less successful.

If, upon introducing the needles into the tissues, they are put in connexion with the conductors of a voltaic pile in action, the patient experiences a more pungent pain; but after some time, it diminishes, when the pile has ceased to act, and then commonly disappears.

But let us leave, for the present, these various opinions, and continue the description. When it is thought that the needle has remained a sufficient time, which varies very much, according to the affection, it is withdrawn. At first, M. Cloquet withdrew the needles as soon as the original pain, and that produced by the instrument, had ceased. But, at present, experience has taught him, that it is much better to let them remain a longer time. Accordingly, he has allowed them to remain eight, ten, and twenty-four-hours; and, finally, in many diseases, forty-eight hours. Afterwards, he withdraws them, very often to be replaced by others.

The extraction is generally more painful than the introduction; especially if the needle has remained a long time, or has been charged with the electric fluid by aid of the pile. No blood is drawn; yet sometimes I have seen one, and even many small drops, ooze out. If we now examine the needle, it has undergone a very evident change; its colour is no longer the same. If it has remained but a short time, its point only is oxidized: the remainder of the surface which had been in contact with the tissues, presents concentric layers of oxidation, separated by spaces, in which the natural colour is still exhibited; But if it has been inserted for a long time, the oxidation takes place over its whole surface; and this will be much more considerable, if the pile be employed. The nature of the tissues, and the intensity of the pain appear to create some difference. In proportion as the part is more painful, or naturally endowed with greater sensibility, so will the oxidizement be more evident.

This oxidizement explains the cause of the painful sensation experienced when the needles are withdrawn; for, having lost their polish, they irritate the parts, already very sensible, and whose sensibility is still further increased by their presence.

If we wish to employ the same needles again, it is necessary to rub them well with emery paper. By this means, the oxide is rubbed off, and the polish restored.

Having performed acupuncturation on the dead body, I allowed the needle to remain about an hour. Upon being withdrawn, it was not in the least oxidized. I afterwards heated the part in which I inserted the needle; it was now oxidized, but very slightly.

Such then are the phenomena, which I have remarked, or which are to be observed, during the operation of acupuncturation. Let us now pass to the examination of the results obtained by its employment.

Part III

This part of my dissertation having for its purpose to make known the circumstances, under which it is proper to employ acupuncturation, as well as to indicate its advantages and accidents, I have thought, that I could not better fulfil it than by a statement of facts; for, *facta potentiora verbis*.

Cephalalgia

Case I

A —, a young man, twenty-five years of age, very irritable, was affected, on the 24th of January, with strong mental agitations. Obliged to make several journeys during wet and unwholesome weather, he found himself considerably fatigued, and felt breaking pains in the limbs, and a very violent pain in the head, especially over the frontal region. I inserted one needle in this place, which I caused to penetrate obliquely under the skin to the distance of about an inch: the introduction was hardly felt. An hour afterwards, the cephalalgia had disappeared, and nothing remained but a slight sensation of wight. The needle was allowed to remain three hours; upon withdrawing it, a small drop of blood followed. The patient was forthwith completely cured. The cephalalgia has not returned.

I remarked upon withdrawing the needle, that it was completely oxidized.

Case II

I had experienced, for several days, a heavy pain in the head. The pain, having become much more acute, deprived me of sleep. Being desirous of proving the effects of the needle on myself, I introduced one, with a hesitating hand, into the temporal region. I acknowledge, that, notwithstanding the intensity of the original pain, I felt very sensibly that produced by the instrument, which did not penetrate to a greater depth than several lines. A short time after its introduction, I felt a species of quivering towards its point; my head appeared to be less heavy, but the pain did not cease entirely. Nevertheless, I was much relieved; sufficiently so, to enable me to read, or to occupy myself with objects, which fix the attention; but three hours afterwards, the pain returned more violently than at first, and at the end of two days, disappeared, as usually happened, without recourse to acupuncturation.

I attributed this pain to the state of the atmosphere* and to my remaining in a very warm room.

Eight days afterwards, being again attacked with a cephalalgia, even more violent than the first, I resorted again to acupuncturation. The pain was felt more particularly on the sides of the head* I inserted a very fine needle in the left temple, in an oblique direction, to the depth of an inch. The introduction was hardly felt. After the lapse of an hour, the pain had disappeared. The needle was allowed to remain ten hours, and, when withdrawn, was much oxidized.

The cure was permanent; nothing remaining but a slight heaviness, which continued for twenty-four hours.

Hemicrania

Case I

J —, aged 40 years, entered the hospital of St. Louis, on the 18th of December, for insupportable pains, on the right side of the head, extending to the ear on the same side. Formerly affected with syphilis, of which she had been cured, the patient has had on the forehead, two nodes, which had suppurated, and of which the cicatrices were still evident. To these nodes, succeeded the hemicrania, of which we are speaking. The pains being unceasing, the patient was deprived of sleep, and suffered from derangements of the digestive functions. Acupuncturation was tried on the 6th of January. The needle was inserted obliquely to the depth of 8 or 10 lines in the temple of the affected side. Being left for two hours, it was withdrawn very much oxidized. Relief was experienced, as soon as the needle had penetrated the tissues, and continued after it had been withdrawn. The patient slept very well the following night, an occurrence which had not happened for a long time: the next morning, the pain had disappeared. This woman remained until the 14th of January, when, not having experienced any new attack of the complaint, she left the hospital, the cure being considered complete.

Case II

A woman, named Menard, entered the hospital of St. Louis, on the 22d of December, for the purpose of being treated for a venereal affection, which had existed for three years.

She had already undergone many courses of medical treatment, but they had all been incomplete.

The patient had experienced for several days past, violent pains in the head, especially on the left side, (pains which she had often felt before,) attended with want of sleep, loss of appetite, inquietudes, heat in the skin of the temple and forehead, and acceleration of pulse.

On the 5th of January, the operation of acupuncturation was performed. The needle was inserted obliquely into the temple on the affected side, to the depth of about an inch, and allowed to remain about two hours. From the moment the needle traversed the tissues, the patient ceased to suffer.

The pain had not returned up to the 22d of January.

Neuralgia

Case I

The subject of this case was a young man, strong and of good constitution. He had experienced for many years, very acute pains in the superior part of the head. They were superficial, and seemed to exist between the hairy scalp and the bone. The least touch on the hair produced a painful sensation. The patient was unable to assign any cause for the affection. Having consulted M. Cloquet about the end of December, a needle was placed upon the painful spot, and introduced obliquely under the hairy scalp, where it was allowed to remain for an hour. Almost as soon as it had penetrated, the pain ceased, and nothing remained but a heaviness. The next morning, the acupuncture was repeated. Although the pain had not returned, he still had this time some relief; nevertheless the heaviness, of which we have spoken above, still continued. The individual, considering himself cured, did not return. Fifteen days afterwards, the original symptoms reappeared, but with less force. After having suffered for fifteen days, without employing any remedy, he returned. A needle was now placed as before, and allowed to remain for twenty-four hours. The success was not less striking than in the first instance. For five days, the acupuncturation was continued, the needle being changed every day. From this time, all symptoms disappeared, the young man experiencing no other suffering than that produced by the needle.

The cure appeared to be complete on the 28th of January.

Case II

H —, aged about twenty-six years, a domestic, living near the Bicêtre, had suffered, for two months, from very acute pains along the course of the sciatic nerve. These pains extended to the leg, where they were equally severe, and followed the direction of the fibular nerve. This man, having presented himself at the consultation of the Hôtel-Dieu, three needles were inserted, one in the upper and external part of the thigh, another, two-thirds down the thigh, and the third, in the external part of the calf of the leg. Their introduction appeared to cause much suffering. Shortly afterwards, he felt shooting pains, very acute prickings, and dragging sensations, proceeding from all parts of the limb, and converging to the points of the needles. In a quarter of an hour, he was affected with syncope, but soon recovered.

The needles were allowed to remain three-quarters of an hour, after which, they were removed. The patient experienced no relief, and went away in the same state as when he came, except that he was very weak. This patient has not returned.

Case III

Peter Oudia, about forty-seven years of age, formerly a soldier, strong and robust, never having had any other diseases, but wounds, and a venereal affection, which was well treated and completely cured — was affected, in September, 1821, with very acute pains, especially during the night. They were seated in the right inferior extremity, extending from the sacrum to the toes, following always the course of the sciatic nerve. The patient experienced a very evident sensation of cold in the limb. The pains were constant, attended with exacerbations, and very violent cramps. He attributed the disease to fatigue and exposure to cold and wet. Liniments of oil with ammonia, had been prescribed in the beginning, but without success. Not being able to walk, or to supply his wants by labour, he entered the Hotel- Dieu. Two bleedings in the foot, and two in the arm, were practised, very little blood being drawn each time. Tepid baths were employed as co-operating remedies; but their effects were not advantageous, being followed by no abatement of his pains. This method of treatment was laid aside, and recourse was had to blisters, applied in various places, along the course of the sciatic nerves; but without good effect. The pain became even more violent. Not only the patient could not walk, but lie experienced more acute pains than before.

It was now determined to try whether the extract of mix vomica would succeed better. It was given perseveringly in large doses, but the result was far from corresponding with the benefit expected. The patient had convulsions and vomitings, and was brought to a state analogous to that caused by drunkenness, being stupid, with heaviness of head: but without any alleviation of the original pain. Having quitted the Hôtel-Dieu, he placed himself in the hands of empirics. A woman induced him to take fifty-two purges in sixty days. His frame proved sufficiently robust to withstand so incendiary a course of treatment. By his account, he was a little relieved; but shortly this supposed relief ceased to be perceived, and he was obliged to return to the hospital. Galvanism was now tried; but his condition was not improved. Finally, having come to St. Louis, two needles were inserted, over the sacrum, on the 5th of January, 1825. They penetrated about an inch, and were allowed to remain an hour. The pain disappeared, and afterwards returned in the knee. Five days afterwards, two other needles were inserted obliquely into the sides of this articulation, and allowed to remain two hours: the pain disappeared again; but shortly returned in the ankle joint. This part was then treated as the knee, two needles being introduced, and allowed to remain three hours. As soon as they were inserted, the pain disappeared.

The patient having still felt slight attacks for several days, five needles were successively introduced in the direction of the sciatic nerve. Since the 24th, he has felt nothing but a slight trembling; he sleeps very well, and begins to walk, although with great difficulty. There is, nevertheless, reason to believe, that exercise will be sufficient to restore him to the use of his limbs.

Muscular rheumatism

Case I

The young man, who is the subject of this case, is fifteen years of age, very lively, strong, and robust. He had suffered under very acute pains in the back part of the neck for nearly eight days. Pressure on this part was painful. He could neither turn his head to the right or left, nor raise or depress it without suffering.

A needle was placed upon the painful point, and introduced to the depth of an inch. The pain produced by the acupuncture was far from being acute. Ten minutes afterwards, the areola formed, and the pains disappeared. The needle was removed at the end of three-quarters of an hour. The pains have not reappeared. Having examined this young man, the morning after the operation, I saw him perform all the movements, which it was impossible to execute the night before j and nothing remained of the affection but a slight stiffness. (Consultation of the hospital St. Louis.)

Case II

M —, a student, aged about twenty-four years, strong and robust, suffered from very acute pains in the posterior part of the neck. These were not continual, but occurred when the weather was cold, afflicting the patient for three or four days, sometimes even for eight, and afterwards disappearing without treatment. When present, they are sufficiently acute to prevent sleep, and constrain considerably the movements of the head. Pressure is but slightly painful. At a moment when the pains were strong, I placed a needle an inch above the spinous processes of the cervical vertebrae, and inserted it to the depth of about an inch. The introduction was very painful. While it remained in, the patient complained of very acute pricking sensations at the place of its insertion. The areola appeared at the end of fifteen minutes. The needle was withdrawn in forty minutes, its removal being not less painful than its insertion.

The pain ceased, but returned in the course of several hours. The patient, thinking that he had paid too dear, for the momentary relief obtained, in the pain caused by the operation, was unwilling to resort to it a second time. The pain continued. This case was recorded in December.

Case III

A man, strong and robust, 28 years of age, entered the Hôtel-Dieu, to be treated for very acute pains, situated in the deltoid muscle, and which were propagated to the muscles of the arm. The pains were constant, without any general inflammatory action, and had existed for the last three weeks. Professor *Récamier*, persuaded that acupuncturation was suited to this case, performed the operation. Two needles were inserted, one in the centre of the deltoid, the other towards the external edge of the triceps, to the depth of about an inch. The introduction was but slightly painful. Almost immediately afterwards, the patient was entirely free from pain. The needles were withdrawn, after remaining inserted three-quarters of an hour. The pain has not made its appearance since.

Eight days afterwards, the patient left the hospital, being considered cured.

Case IV

A woman named Ranci, forty-eight years of age, a patient at the Hôtel-Dieu, had been afflicted, during the course of her life, with a great number of diseases, and for the last three years had suffered from very acute erratic pains. These had been treated by bleedings, both general and local, cups, baths, fumigations, and uniformly without success. The pains were felt strongly in the direction of the vertebral column. Two needles were placed in the muscles of this part; and, being allowed to remain about an hour, no relief whatsoever was obtained. The operation was not, at this time, repeated. But fifteen days afterwards, the pain having been translated to the deltoid region, a needle was inserted towards the inferior angle of the muscle of that name. It was allowed to remain an hour, and, upon being withdrawn, the pains had become more acute. From that time, acupuncturation was abandoned.

Case V

A washerwoman, named Giraud, forty-four years of age, entered the hospital of St. Louis, on the 9th of November, for the purpose of being treated for a rheumatism, seated in the muscles of the arm. The pain was felt more especially towards the point of the deltoid. This pain, which was very acute, had existed for eight months, and had resisted local bleedings, as well as cataplasms, vapour and water baths, and blisters. No evident relief was even obtained. On the 3d of January, a needle was inserted in the middle of the deltoid, to the depth of about an inch, being slightly inclined from below upwards. It was allowed to remain two hours. Whilst it remained inserted, the pain was less, and the relief continued after it was removed, the patient being able to move her arm, what she had not been able to do for a long time. Six new acupunctures were afterwards performed on different parts of the arm. After the third operation, the pain had left the rest of the arm, and become seated exclusively in the deltoid. The two last acupunctures were performed on this part, without any relief. The needles were then inserted for eight hours, after which the patient was allowed to rest for eight days. On the 21st, a new operation was performed, the needle being allowed to remain thirty-six hours. Some relief was experienced, but there was not obtained the success, which the first trials seemed to promise. Nevertheless, the repetition of the operation gives reason to hope for a favourable result.

Case VI

In the latter part of December, a young officer, belonging to the garrison of Paris, came to the consultation of the hospital of St. Louis, to solicit relief for a very acute pain, which he felt in the right thigh and leg. The pain, which had existed for several months, had, without question, been induced by exposure to cold and humidity. It was seated in the external part of the thigh, and extended to the calf of the leg. The officer walked with much difficulty. A needle was inserted in the place where the pain was felt most acutely. The introduction was not very painful, and the areola was a long time in forming. A quarter of an hour after it was inserted, he felt a very intense pain in the spot occupied by the instrument, and shootings and tremblings, proceeding from the calf of the leg towards the point of the needle. Shortly afterwards, the uneasiness became general, and the face pale, and finally the patient fainted. The needle being immediately withdrawn, cold water was thrown on the face, and the patient exposed to a very cold air, and shaken, to recall the sensibility and circulation. He remained more than twenty minutes in a state intermediate between syncope, and his natural condition. Finally, we succeeded in bringing him to himself, and rendering him sufficiently strong to enable him, after the lapse of an hour, to regain his barracks on foot. The operation relieved him but very little; for, although his pain was a little diminished, yet I conceive, the diminution ought rather to be attributed to a modification of sensibility, caused by the syncope that occurred. In the course of ten days, the pains had disappeared gradually, and the patient found himself entirely well.

Case VII

A married woman, forty-one years of age, strong and robust, had suffered for about two years from pain, seated in the muscles of the back part of the thigh, and those of the leg. The pains were nearly constant, depriving her of sleep, and equally severe both night and day. Atmospherical changes exerted a great influence. When the weather is tempestuous and moist, she suffers most, and is unable to walk without a stick; and even now she does not walk without much difficulty. The affection had been combatted by local bleedings, cups, and baths, but all without success. I proposed to her, on the 6th of January, the day on which I saw her for the first time, to make trial of a slight operation. My proposal being acceded to, I inserted a large sized needle in the superior and back part of the calf of the leg, burying it perpendicularly to the depth of about an inch and a half. Notwithstanding the size of the needle, its introduction was but slightly painful. The areola was very soon formed. The patient experienced shooting pains, and a sensation of drawing seemed to proceed from the thigh and foot, directed towards the point of the needle. The original pain had already disappeared, hut a fainting fit supervened, attended with paleness, general debility, dejection* and diminution of pulse. The loss of sense was not complete, and a few drops of fresh water, sprinkled on the face, removed this slight accident, of which, in a short time, no trace remained. The needle having remained three-quarters of an hour, I withdrew it, the point being very much oxidized. The patient no longer felt any pain; she rose from her seat, and, at first, was afraid to support herself upon the limb, which had been the seat of the disease. No longer perceiving any pain, she laid down her stick, and walked with much ease. Nothing remained hut a slight stiffness in the part. Having requested the patient to return, if the pain should reappear, and not having seen her since, I am led to believe that she is entirely cured.

Case VIII

A woman, named Pion, seventeen years of age, a patient at the Hôtel-Dieu, delivered eight months before, had been affected with an inflammation of the abdominal viscera, which was treated by the frequent application of leeches. The patient left the hospital two months after her admission, imperfectly cured, still experiencing constant pains in the abdomen. These pains having become more acute, she returned to the Hôtel-Dieu, in the following state: — tongue slightly red at its* point; respiration painful upon taking a deep breath; belly swelled, and painful on pressure; appetite not impaired; pains increased by abstinence; digestion easy for every species of food; thirst moderate; bowels affected with looseness. The patient was put upon gummous drinks. During the first days of January, the patient having been seized with a very acute pain in the shoulder and left arm, sixty leeches were applied, without procuring any relief. Several days afterwards, the pain continuing, recourse was had to acupuncturation. The operation was put in practice twice, at intervals of four days, two needles being employed each time. The first were placed, one near the point of the deltoid, the other, two inches below it, both being inserted to the depth of about an inch. The introduction appeared to be painful. The patient experienced a numbness in the arm, and a sensation of heat near the point of the needles. Five minutes afterwards, a very evident relief was obtained, which lasted forty-eight hours. After this time, the pain returned, but with less intensity. Two other needles were now inserted, one on the inside of the arm, the other on the outside, but to a less depth than the first.

The pain was this time almost entirely removed. Some hours afterwards, it disappeared completely, and continued absent for six days; at the end of which time, it began to be felt again, but in a very slight degree. Its intensity having increased a little, a blister w r as applied to the arm, but it gave no relief. The first needles were allowed to remain inserted for two hours.

The abdominal pains having increased, it was decided, on the second of February, to insert two needles in the iliac regions. They were made to penetrate to the depth of an inch and a half. Immediately afterwards the pain ceased. The needles were withdrawn, after having been allowed to remain seven hours. Their insertion and withdrawal appeared to be equally painful. Since this time, the patient has felt no pain, and pressure, heretofore so painful, produces to-day, (4th Feb.) not the least uneasy sensation.

The looseness has neither been increased nor diminished. During the time the needles remained inserted, the patient experienced some slight feverish symptoms.

Case IX

A woman, named Dumaroi, an upholsterer, fifty years of age, entered the hospital of St. Louis, on the 18th of November, to be treated for rheumatic pains. This woman, forced by her condition to work in cold and moist situations, after having past a night in the vast apartments of the Tuilleries, was seized with pains, which invaded successively the different parts of the body. They soon became fixed in the right leg, afterwards in the thigh, and finally in the buttock of the same side. Moderately painful at first, they afterwards increased to such a degree as to render walking impossible. The remedies first employed, were a blister, and camphorated liniments, but without effect. Fumigations were then had recourse to, but, far from obtaining any mitigation, the pain became more acute. Sedlitz water was administered internally. When the patient laid down, the pain disappeared. In the last half of January, acupuncturation was tried. Three needles were inserted at the same time, to the depth of about an inch; one near the ankle, another in the thigh, and the third in the middle of the buttock. Almost immediately, there arose around the first, a button-shaped eruption. The needles were allowed to remain for three days. The pain disappeared twelve hours after their introduction, the patient feeling nothing in the limb, except a sensation of stiffness and constraint.

Eight days afterwards, having experienced a new attack of her disease, but in a different place, (the loins), two other needles were introduced, one in the lumbar region, and the other under the great trochanter. An eruption occurred round one of the needles, as in the first instance. The pain was removed, as after the first acupuncturation, and at the present moment, walking causes only a slight feeling of fatigue.

The patient has left the hospital cured.

Case X

A married woman, named Constance Vial, thirty-six years of age, had always enjoyed good health, and been regular in the catamenia, when, within the last two months, she has felt very acute pains in the soles of the feet. Shortly afterwards, these parts became swelled and very painful. The swelling and pains invaded successively the ankle, knee, and hip joints. The pain was so severe in the last mentioned part, that it became necessary to apply leeches. It disappeared in this part, only to manifest itself still more severely in the other joints mentioned. Emollient cataplasms were now applied, hut produced but little relief. Twenty leeches were then applied upon each of the affected limbs, being distributed on the joints. Twenty-four hours afterwards, the patient felt nothing but a little numbness* and was able to walk. In the course of a day, this numbness ceased, but the hands became affected with swellings, which extended to the arms. On the 15th of January, a new application of twenty-five leeches was prescribed, which, on this occasion, produced no relief. On Tuesday, the 25th, the patient having been sent by M. Magendie, acupuncturation was employed. At this time, her hands were very much swelled over their whole surface. The contact of any body caused her to put forth cries. Upon making pressure, the impression of the finger remained. Two needles were introduced obliquely, between the skin, and the ligaments found at the back part of the wrist joint. Their introduction was hardly felt. Being allowed to remain about two hours, no immediate effects were produced. Nothing was observed but a diminution in the swelling. The next day, the 26th, the pain ceased. On the 27th, acupuncturation was again employed, for the same space of time. From this moment, the patient felt nothing but a little numbness, and even this went off the next day. At present, she has recovered the free use of her hands.

Case XI

Curtel, twenty-seven years of age, a servant, strong and robust, after having fatigued himself a great deal, had a swelling of the knee, attended from the first with very acute pain. Obliged to walk, notwithstanding the pain, for the last six months, the swelling increased, and the lameness became very considerable. It was especially above the patella, that the pain was seated. He remained for four months, without doing any thing, only avoiding to walk as much as possible. The progress of the malady forced him to enter the hospital of St. Louis, where he was admitted on the 22d of December, 1824. As soon as he came in, without having recourse to other means, acupuncturation was employed. Two needles were inserted, and made to penetrate obliquely on the sides of the patella. The patient experienced no pain on their introduction. Being allowed to remain from an hour and a half to two hours, they were withdrawn, and the pain and swelling found to be a little diminished. The operation was repeated five times in five days, the needles being always allowed to remain for the same period. The patient experienced some relief each time, but less than after the first operation. The ward in which he was, having changed physician, the needle was no longer employed. Blisters were now resorted to for eight days, but without any benefit. The patient felt a strong desire, that recourse should again be had to acupuncturation.

Spontaneous luxation

Case

Autefeuille, eighteen years of age, born at Melun, and bearing the marks indicative of health, resided in a moist situation. She had commenced, about two years before, to experience very acute pains in the left hip joint. The complaint continuing constantly to increase, the young woman, no longer able to walk, and obliged to keep her bed, had recourse to the advice of physicians. Leeches, cataplasms to the painful part, and emollient baths, were tried in vain; for nothing appeared to arrest the progress of the disease. The limb first lengthened, and afterwards grew shorter, from the occurrence of its luxation. The head of the femur was carried upwards and outwards, and, acting as a foreign body in its new situation, caused a violent inflammation. Being combatted by the remedies just indicated, it at first yielded a little; but afterwards the patient had no respite from her sufferings. Admitted into the hospital of St. Louis, now four months, cups were frequently applied, but with little relief; and the antiscrofulous treatment, with rest and cataplasms, was prescribed. In spite of all these means, the pain continued, when, on the 15th of December, about six weeks ago, acupuncturation was tried, behind, and about an inch under, the great trochanter. In ten minutes, the areola formed, and the pain disappeared. The needle was allowed to remain about two hours; after which, upon its being withdrawn, it showed traces of oxidizement. The pain, which had ceased during the operation, did not return for several days. The operation was repeated five times, in five days. The patient has not experienced any attacks of the original pain, except within the last fifteen days; and they are much less sharp, and only manifested, when the attempt is made to walk, and even then they are hardly to be felt

Contusion of the pectoral muscles with haemoptysis

Case

A man, above sixty years of age, strong and robust, having a strongly developed muscular system, was struck and thrown down. His body presented contusions in several places, but pain was felt most severely on the right side of the breast. He breathed with the greatest difficulty, and expectorated blood in considerable quantities. A physician, called in, prescribed twelve leeches upon the right side of the thorax, under the arm. The patient was relieved, and ceased to spit blood, but continued to have a very great difficulty in breathing. Wishing to obtain relief, he applied to the consultation of the hospital of St. Louis, fifteen days after the accident. On the 20th of December, he was in the following state: — pungent pain in the right side; respiration painful, short, and frequent, the ribs scarcely rising or falling; cough slight, and expectoration free from blood; pressure on the internal extremity of the pectoralis major is attended with pain; pulse, strong, full, and regular; skin free from heat. A very long needle was inserted under the nipple, over the most painful point, and made to penetrate nearly two inches, so as to reach the pleura, or even the lung itself. From the moment the needle was inserted, the relief was manifest; the patient ceased to suffer, became easier in his respiration, and, to avail myself of his own expression, we had relieved him of a weight. Pressure produced no further pain. The needle, after remaining three-quarters of an hour, was withdrawn. The patient was merely relieved by the first operation; but six other acupunctures, performed in the space of ten days, freed him entirely from the complaint. It is here worthy of remark, as is most generally the case, that the effect of the operation was not subsequently so prominent, as on its first trial.

Pleuritic pain with haemoptysis

Case

B —, twenty-two years of age, of a lymphatic temperament, married about three years, and mother of a healthy child, occupying a damp ground floor, was taken with a swelling in the ganglions of the neck. The engorgement was neither large nor inflammatory. Six months before, this woman had been attacked with pretty sharp pains in the knee joint. The part shortly after swelled, and became red, and the pain increasing, she found it impossible to walk. Twenty leeches were applied, and emollient cataplasms prescribed. These remedies gave some relief, and the glands of the neck disappeared in the course of several days, without any obvious cause. Almost immediately afterwards, sharp pains were experienced in the right side of the chest, attended with short and difficult respiration. Pressure upon this part of the thorax was attended with much pain, and the patient could not lie upon the affected side. The patient had cough attended with pain, and the expectoration was streaked with blood. The skin was hot, and the pulse accelerated. Twenty-four hours after the appearance of the first symptoms, acupuncturation was put in practice, by inserting, perpendicularly, a single needle, under the right nipple, to the depth of fifteen lines. The introduction was hardly felt. As soon as the needle was inserted, the pain diminished. In five minutes, the areola formed, after which the pain disappeared, and freedom of respiration was reestablished. After remaining a quarter of an hour, the needle was withdrawn without pain to the patient, and found but little oxidized, and that only near its point. In half an hour after the withdrawal of the instrument, the pains returned, but only to disappear at the end of five minutes. Subsequently, there was no spitting of blood, no difficulty of breathing, in short no symptom of remaining disease. Mucilaginous drinks were [inscribed, and a light diet. The cure was immediate, and the disease did not return. On the same day, this woman, so afflicted before the operation, returned to her occupations. The relief was so prompt, that, naturally credulous, she supposed it to be the effect of witchcraft.

Chronic ophthalmia

Case I

Vitala, an embroiderer, forty-two years of age, has always been regular in the catamenia, and enjoyed good health. For about five years, she has had a constant weeping from the eyes, and a slight irritation at the edge of the eye-lids; when, about ten months ago, she was affected with acute ophthalmia, which passed into a chronic state, liable to be converted into the acute form, under the influence of the slightest exciting causes. Since this time, the disease having progressed, the eye-lids have become red, everted, and deprived, for the most part, of the eye-lashes, and the conjunctiva very much inflamed, the impression of light being extremely painful. Having presented herself at the hospital of St. Louis for advice, about ten days ago, (15th of

January,) two needles were inserted, one in each temple, and allowed to remain for twenty-four hours, being, at the end of that time, immediately replaced by others. This plan was pursued for four days, after which the needles were allowed to remain forty-eight hours. From the commencement of the treatment, the disease has always been on the decline, and, at the present day, the inflammation has almost entirely disappeared, the impression of light being no longer painful. This woman has not interrupted her daily labours, or employed any other curatire means.

Case II

A man, named Schneider, twenty-two years of age, of a lymphatic temperament, following at present the trade of a shoemaker, but formerly that of painter on porcelain, has been affected with an ophthalmia, which has existed for more than four years. The edge and internal surface of the eye-lids are red, and the vessels of the globe of the eye, very much injected; while the impression of light and shining bodies is painful. The slightest deviation in diet is sufficient to re-excite the state of acute inflammation. Treated repeatedly with resolutive collyria, the intensity of the symptoms, by these means, was diminished, without effecting a complete cure. Having presented himself, on the 22d of January, at the consultation of M. Cloquet, a needle w r as inserted obliquely in each temple, to the depth of nine lines, and allowed to remain for twenty-four hours, after which they were withdrawn. The inflammation was found sensibly diminished. Two other needles were made to replace the first, and allowed to remain for the same space of time. The affection, making rapid improvement as the cure approached, recourse was had exclusively to prolonged acupuncturation. In the fifth operation, the needles were allowed to remain forty-eight hours, and to-day, (January 29th,) the cure is almost complete. During the course of the treatment, no accident occurred; the patient having felt nothing, except a slight itching towards the internal angle of the eye, in the same spot, in which the inflammation began to abate. The presence of the needles was so slightly inconvenient, that Schneider has not, for a moment, been taken off from his habitual occupations.

Case III

A woman, named Frances-Henrietta, twenty years of age, had contracted a running, two years before, the consequence of an impure connexion. It had continued for two months, and then disappeared, during the administration of astringent drinks, by an empyric. For six months, this woman remained, to all appearance, completely cured. But after this time, she felt the most violent pains in the whole of the left side of the head, and in the corresponding eye. These were constant, totally depriving her of sleep, and causing the most frightful agitations. Two months afterwards, she felt a very acute irritation in the left cheek; and shortly after, this part was covered with a crust, resulting from the concretion of the fluid, which exuded from the irritated spot. The affection made very rapid progress, and proceeded, in the course of a few days, to affect the eye-lids. At this time, she was admitted into the hospital of St. Louis, being in the following state: — her face was disfigured by the herpetic eruption, of which we have spoken. The scab was entirely surrounded with a circle of a purplish-red colour. The left eye, very much inflamed, could not support the contact of light, and exuded a very abundant puriform matter, more especially during the night. The pains were felt over the whole anterior part of the head. They were constant, hut much more acute during the night than the day, and prevented the patient from tasting the sweets of sleep. To combat this affection, general and local bleedings, blisters, a seton, pediluvia, and purgatives were all employed without effect. Two anti-venereal courses were resorted to, but without the least amelioration to the patient. Wishing to try the effect of acupuncturation in so obstinate a pain, M. Julius Cloquet inserted a needle in the temple of the affected side, and allowed it to remain for an hour. The pain disappeared entirely, and from that moment, the patient was enabled to resign herself to sleep, what had not occurred to her before, unless when worn out by fatigue and watchings. The pain reappeared fourteen or fifteen hours afterwards, but in a less degree. The second day after the operation, a new one was performed. The needle was allowed to remain two hours; and the relief was not less marked, the pain disappearing entirely. Since then, she has felt it anew, but much less severely. During the space of a month, the operation was not repeated; but after this time, on the 15th of January, *persistent acupuncturation* was resorted to. The needle was allowed to remain, being renewed every three or four days. The eye is now able to support the lights there is no more redness or puriform secretion; and the tetter itself is found to improve very much, and contracts considerably. Doubtless acupuncturation cannot remove the existing venereal affection; but, in the present instance, it has the merit not only of arresting its progress, but also of diminishing the local affection. (At the present moment, (12th of February,) the tetter has completely disappeared, and the patient experiences no pain.)

Periodical haematemesis

Case

A woman, named Duperier, twenty-nine years of age, having been affected about nine years before, with a very violent moral commotion, immediately after the flowing of her courses, was suddenly attacked with hysteria. From this time, the menses have not appeared; but they are replaced by a periodical vomiting of blood, lasting generally three or four days, and accompanied with very sharp pains in the epigastrium.

Antispasmodics, and preparations of cinchona, had been used, but without affording relief. The only measure that has improved the symptoms, was the employment of sinapised pediluvia.

Admitted into the Hôtel-Dieu on the 29th of January, the periodical vomiting appeared on the 31st. Twelve leeches were prescribed to the vulva, and forty to the epigastrium; but the vomiting continued, and the pain of the epigastric region underwent no mitigation. On the 2d of February, acupuncturation was put in practice. Two needles were inserted, one in the epigastrium, the other near the umbilicus, to the depth of an inch. Immediately upon their introduction, the pain and vomiting ceased. They were allowed to remain two hours. An hour after they were withdrawn, the vomiting reappeared, but the habitual pain was considerably diminished. This patient found herself much better.

Uterine pains

Case I

A woman, named Dujardin, twenty-seven years of age, strong and robust, having contracted a running eighteen months before, had allowed it to proceed for seven months without any treatment. Aware of her disease, and driven to despair, she determined to destroy herself, and, seizing a razor, inflicted a dangerous wound on herself in the neck. Carried to the Hôtel-Dieu, she was cured by the attentions of the skilful surgeon of that hospital. Since then, she has had various affections, which caused her to remain. The treatment of these affections not coming within the province of surgery, she was transferred to the wards of M. Récamier. It was then that she made known, now six months since, the cause that led her to commit so reprehensible an act of despair. Being affected with syphilitic ulcerations in the internal part of the labia, the anti- venereal treatment was prescribed and continued for three months. The chancres were cured, but the running continued. Astringent injections were then tried for three weeks, and afterwards baths of the vapour of cinnabar: the running now stopped; but, upon the baths being suspended, it reappeared as abundant as at first, staining the linen. At this time, the patient began to experience very acute pains in the region of the womb, attended with shootings, prickings, and a sensation of continual burning. Her sleep was disturbed. She always had a febrile disposition towards evening; her digestion was bad, and, during the process, her pain was more acute. Towards the latter end of December, a long needle was inserted, to the depth of two inches at least, in the sides of the vagina. The patient scarcely felt the insertion of the instrument, which was allowed to remain for an hour and a half. She did not experience any relief while it remained. The pain became even more acute upon the withdrawal of the instrument, which was much oxidized; but shortly, it diminished gradually, and at last disappeared. The patient was for six days entirely free from pain; but the discharge continued. After that time, the pain returned, but less violently than at first. The operation has not been repeated, the plan of treatment by cicuta and baths of the vapour of cinnabar being preferred. This new treatment has relieved her only in a slight degree.

Case II

A woman, named Adelaide Colin, thirty-five years of age, following the occupation of a cook, had enjoyed, until the age of 32, a good state of health, notwithstanding her first menstrual discharge had been established with difficulty, and she had experienced much uncomfortable feeling. From this time, the catamenia had always been regular and abundant, announced, however, by abdominal pains, but not sufficiently strong to disturb the functions. Having experienced a violent fright three years before, at a time when the menstrual flux existed, it became suppressed instantaneously. Soon after, shooting pains made their appearance in the womb, propagated to the hypochondriac regions, and the abdomen swelled. The catamenia have not since reappeared. This girl was attacked, several months afterwards, with hysteria. A volume would not suffice to detail her disease from this period. She had vomitings of blood, a diarrhoea which lasted several months, a very violent pneumony, very strong pains in the head, which afterwards became directed towards the back; but the point, to which a very sharp and constant pain was always directed, was the womb. All the other symptoms appeared to originate from this organ: all the functions were deranged. Her state, for three years, has never been the same for eight days successively. A single symptom only was constant; namely, the uterine pain; and yet repeated attentive examinations discovered no lesion of the uterus. This woman, whose case I followed for more than a year, at St. Louis, in the wards of M. Lugol, entered successively at St. Antoine, and the Hôtel-Dieu, where she has been for the last six months, in the wards of M. Recamier. The reader may form some idea of the complication of this case, by noting the different treatments to which the patient was subjected. At St. Louis, general and local bleedings, cold baths, sixty and more spout-baths, antispasmodics of all kinds; tonics, purgatives; emetic tartar, in doses of ten and twelve grains, repeated frequently; but every thing without success. Since she has been at the Hôtel-Dieu, thirty-three bleedings have been practised; she has undergone a mercurial treatment to the extent of salivation, suffered six moxas to the internal parts of the labia, and two on the dorsal region, used the caustic, preparations of cicuta, and many other remedies; but all without any useful result. Certain local symptoms were diminished, but the catamenia were never made to appear; neither was the uterine pain mitigated, a pain so acute, that it was characterised by the patient as resembling the constant *gnawing of a dog*, tearing the tissues. Towards the end of December, acupuncturation was put in practice. Three long needles were inserted into the sides of the vagina, to the depth of about one and a half inches, and allowed to remain an hour and a half. The introduction was by no means painful. The needles were no sooner introduced, than the pain which existed in the uterus, and thence extended over the whole of the abdomen, loins, thighs, and even the knees, ceased immediately. Upon withdrawing them, they were found to be entirely oxidized, and covered with rust. An unusual oozing of serosity from the sides of the vagina was now produced, all painful sensation disappeared, and the patient enjoyed a tranquil and refreshing sleep, an occurrence which had not taken place for a long time; for previously, she had slept for some moments only, when worn out by pain and watching. She regained her appetite and strength, was enabled to get up, and became gay; in short, her whole existence was changed. She remained for six days in this satisfactory state. It is proper to remark that acupuncturation was not repeated. After this time, the uterine pains began again to be felt. Moderate at first, they gradually increased, but never regained their former intensity. The treatment was afterwards confined to preparations of cicuta, when, on the first of February, the patient had a considerable difficulty of breathing, a general oedema of the body and thighs, a copious vomiting of pure blood, and, in short, was in such a state as indicated the near approach of death. On the morning of the 2d, three needles were inserted; two below the base of the breast, one on each side, under the cartilaginous margin of the ribs, and the third, below the xiphoid cartilage; all to the depth of at least an inch. The immediate consequences were, cessation of the uterine pain and vomiting, facility in respiration, and a general improvement in feeling. The introduction of the needles was hardly felt. After having remained eight hours, they were withdrawn in an oxidized state. The next day, the 3d, I saw the patient, gay and pleased, having slept well, and feeling herself altogether in a satisfactory state. The oedema had not disappeared, and there remained some degree of tenderness in the abdomen. Two other needles were inserted as before, but upon the sides of the umbilicus. Being left for the same space of time, they were withdrawn, very much oxidized. The abdominal pain ceased to be felt, and nothing remained, but a feeling of dragging in the groins. No doubt, care will be taken not to neglect the use of the needle for the future.

Several journals have made mention of a blind young woman, restored to sight by acupuncturation. As a history of this truly interesting case will shortly be published by M. Roger, resident student of M. Alibert, (a young physician, who had for a long time lavished attentions on her,) I shall confine myself to giving merely some notes upon it.

Case III

Miss Lise, nineteen years of age, very irritable, possessing the temperament, called nervous, had, two years before, been brought under the influence of very violent grief, during the time of the menstrual flux. She was then seized with convulsions, which lasted more than an hour, and left her in a diseased state. From this time, her catamenia have not appeared, and her health has undergone a change. She has experienced, successively, various inflammations of the thoracic and abdominal viscera, and suffered very acute and constant pains in the side of the head. It was reasonably conjectured, that, in the midst of this disordered state of all the functions, the uterus, the primitive seat of the disease, did not remain inactive. The patient had felt, almost constantly, acute pains in that part. She was affected successively with spitting and vomiting of blood, abdominal pains, with swelling of the belly, looseness, obstinate headache, and finally pains which pervaded all parts of the body. All these morbid symptoms were accompanied by frequent attacks of hysteria. Notwithstanding the application of seven hundred and eighty leeches, the employment of twelve general bleedings, caustics, sinapisms, pediluvia, and the use of many medicinal preparations, indicated by the various conditions of the patient, the catamenial flux could not be re-established, the suppression of which was, unquestionably, the cause of the train of symptoms, by which the patient was affected. On the 10th of November, upon coming to herself, after an attack of hysteria, she found herself deprived of sight, and plunged in profound darkness; in a word, her blindness was complete. Headaches, more violent than ordinary, were all that was felt during the preceding days. On examination, the globe of the eye presented no appearance of alteration. Many means were tried ineffectually to restore the sight, such as bleedings, laxatives, derivatives to the intestinal canal, and revulsives; but I repeat, every thing was fruitless. The success of acupuncturation multiplying to a great extent, in a vast number of diseases in the hospital of St. Louis, M. Roger had recourse to it. On the 5th of January, three needles were inserted, two in the temples, and the third in the forehead, and allowed to remain an hour and a quarter. The pains were somewhat diminished; but otherwise, no change occurred. On the 24th, the needles were again employed, remaining as in the first instance, and inserted obliquely to the depth of an inch. The effect was still the same. On the same day, two other needles were employed, which were, this time, allowed to remain inserted for twenty-one hours; but no favourable influence on the organs of sight was produced by the operation. Indeed, on this same day, a light was approached to the patient, without its being perceived, until detected by the heat, to which it gave rise. On the 26th, two acupunctures were again practised at nine o'clock in the morning, the needles being withdrawn six hours after. The sight now returned as suddenly as it had disappeared, and without the manifestation of any extraordinary phenomena; and this youthful patient, who had been plunged in darkness for seventy-six days, was, at the instant, astonished at the unexpected sight of surrounding objects. It is impossible to convey an idea of her enthusiasm and delight. Since this time, she has been able to read the smallest print; and at the present moment, I have in my possession, a little journal on the subject of her complaint, in her own hand-writing. The other morbid symptoms, such as pain and swelling of the abdomen, &c. were combatted by acupuncturation, performed on the regions affected. Yesterday, the [6th of February, the catamenia made their appearance, and although her constitution has been undermined, hopes of preserving her life should not be abandoned.

Conclusion

FROM the observations which I have detailed, and especially from the multitude of facts I have witnessed at the hospital of St. Louis, where acupuncturation is practised every day on more than twenty patients, I think I am borne out in the assertion, that this operation possesses great therapeutic powers,* that it not only may, but ought to be employed in rheumatic affections, and neuralgies; that, moreover, it succeeds very well in the treatment of many diseases, belonging to the class of phlegmasia, and that it has been censurable that this curative means has remained so long neglected.

But how does the needle act? Is it by subtracting the electric fluid, or is it by derivation, or revulsion? This is a problem, which I do not attempt to solve, believing it far preferable to doubt, than fall into error. It is at least certain, that it has a powerful modifying influence on the sensibility of the organs. I do not range it, as Vicq-d'Azyr has, after moxa and cups. Its action is altogether different. It possesses nearly all their advantages, without their inconveniences; and, very often, has been known to succeed, after moxas had been used without effect.

Acupuncturation may be easily practised on all parts of the body, and is not painful. The physician, who wishes to employ it, has no occasion to surround himself with an apparatus, the sight of which alarms the sick, who are naturally timid persons; while this is an indispensable attendant on the application of the moxa.

In a word, acupuncturation produces in its consequences, neither pain, accident,[1] nor cicatrix.

At the present moment, when this operation is practised in nearly all the hospitals of Paris, its good effects will be easily verified; and if, every where, the same results have not been obtained, I am well satisfied, that this ought to be attributed to differences in the cases, in which it has been employed, and, perhaps, still more to a want of perseverance on the part of the physicians, who have practised it. M. Julius Cloquet daily obtains very remarkable results, by attacking pains constantly as they reappear, and by allowing the needles to remain in the part affected, for five, six, seven, or eight days, in cases, in which their application for several hours had not a sufficiently marked effect. It is to acupuncturation, performed in this manner, that he has applied the epithet *persistent*.

Observation multiplying to a great extent, the general opinion will soon be settled on the subject of this remedy.

It may be proper to remark, that, if, up to this moment, acupuncturation has been employed, in an exclusive manner, in the treatment of diseases, while its good effects have thus been well demonstrated, we ought not to neglect to associate with it as auxiliaries, those measures, w hose good effects have been verified in common practice. Thus, for example, in chronic ophthalmia, at the same time, that

---

[1] At the hospital of St. Louis, where acupuncturation has been practised upon more than four hundred patients, not a single accident has occurred, with the exception of syncopes, of which I have already spoken. M. Cloquet considers, according to his observations, that about a thirtieth of the patients are affected by them.

we allow the needles to remain in the temporal regions, we ought not to neglect the use of resolutive collyria.

What I have here said in regard to ophthalmia is applicable to a great variety of diseases.

But what are the accidents that acupuncturation may cause? I have only observed one, and as yet it is rare; namely, the faintings, of which instances are given in the foregoing cases. As to the pain, more or less acute, which succeeds or attends the operation, it cannot reasonably be called an accident. Some, however, have been more vaguely mentioned. But it is probable, that the authors of these statements have not ascertained their truth by personal observation, as I have reason to believe on sufficient evidence.

If I had designed merely to sound the praises of acupuncturation, I should have given a series of cases all crowned with success, and, without doubt, I would not have been deficient in facts; but my object has been to present a picture, which might be the representation of truth.

Quidquid verum et boiuim rogo.

4. Acupuncture – Elliotson, 1832

.

John Elliotson - Acupuncture [1832]

The Cyclopædia of Practical Medicine 1832, 1:54-57

ACUPUNCTURE. The passing a needle into the body is termed *acupuncture*. From forgetting that the word puncture has two significations, — that it is used to signify both the wound and the act of making it, some have termed the operation *aciipuncturation*. But to subjoin the syllables *ation* to the word puncture or acupuncture, is as improper as to subjoin them to the words preparation or fabrication, each of which already ends in *ation* and has a similar twofold meaning. An exactly parallel error would be to say *manifucturation*.

The most obvious purpose of this operation is to allow the escape of the fluid of oedema or anasarca through the skin, or of the blood when superficially accumulated; but, from an idea that various disorders arose from a kind of subtle and acrid vapour pent up, it was had recourse to, for the purpose of giving this vent, by the Chinese, from time immemorial. From China the practice spread to Corea and Japan, where it has for ages been very common.

Ten Rhyne, (Dissert, de Arthritide, de Acupunctura, &c., London, 1693,) a medical officer in the East India Company's service in 1679, gave the first information to Europe of a practice unknown to the Greeks, Romans, or Arabians; and states that a guard of the Emperor of Japan, appointed to conduct the English to the palace, was seized with violent pain of the abdomen and vomiting, after drinking a quantity of iced water when heated. He took wine and ginger in vain; and then, persuaded that he had wind, had recourse to acupuncture in the presence of Ten Rhyne. It appears that the .Japanese are liable to a violent kind of colic called *senki*, which they regard as too severe to arise from morbid matter in the cavity of the intestines, and ascribe to something morbid in the parietes of the abdomen, the omentum, mesentery, and substance of the intestines, converted by its stay in these parts into a vapour, the escape of which from its narrow prison, by means of acupuncture, is immediately followed by a cessation of the pain and distension. The guard laid himself upon his back, placed the point of a needle upon his abdomen, struck its ! head with a hammer once or twice to make it pass through the skin, rotated it between his fore-finger and thumb till it entered to the depth of an , inch, and then after thirty respirations, as it would appear, withdrew it, and pressed the punctures with his fingers to force out the imaginary vapour. He made four such punctures and was instantly relieved and got well.

The needles are always made of the purest gold or silver, preferably of gold, and well tempered. Their manufacture is a distinct occupation, understood by few, and those few are licensed by the emperor. Some are fine, about four inches in length, with a spiral handle for the purpose of more easily rotating them: and are kept, by means of a ring or a piece of silk thread, in grooves, each capable of holding one, at either side of a hammer, usually made of the polished horn of the wild ox, ivory, ebony or some other hard wood, rather longer than the needle, and having a roundish head covered, on the side which strikes, with j a piece of leather and rendered heavier by a little lead within. Others are of silver only, still finer at their point, but with a short thick handle bent down upon itself; and are kept, several together, in a varnished wooden box lined with cloth: these are not struck with a hammer; but a fine copper canula, about an inch shorter than the needle, is sometimes employed to steady it, and prevent it from entering too far.

The selection of the part fit for the operation, or for the application of the moxa, — the other great remedy of the Japanese, is usually confided to particular persons called *Tensasi*, — touchers or searchers of the parts, while those who apply the needles are styled *Farittate*, — needle-prickers, though occasionally the common people trust to their own experience, taking care only to prick no nerve, tendon, nor considerable blood-vessel. The seat of the cause of the symptoms is the proper part, and delineations of the body are sold conveying this information.

If the patient does not bear the needle well, it is at once withdrawn: but if he does, and the disease proves obstinate, it is introduced two, three, four, five, or six times. The more severe the affection, and the stouter the patient, the deeper must be the puncture.

Kaempfer,[1] a physician who accompanied a Dutch embassy to Japan, in 1691, and again in 1692, informs us that the Japanese make nine punctures, three rows of three each, at about half an inch from each other, over the liver, in cases of colic, and that he himself frequently witnessed the instantaneous cessation of the pain, as if by enchantment.

The orientals do not, however, employ this operation in affections of the abdomen only. In tetanus, convulsions of all kinds, apoplexy, gout, rheumatism, swelled testicle and gonorrhoea, and in fevers both intermittent and continued,

---

[1] Engelbertus Kaempfer, M. D *History of Japan, translated from the High Dutch by Dr. Scheuchzer*. London, 1727.

it is also celebrated among them; enjoying credit, like all remedies of undoubted efficacy in certain diseases, for power which it does not possess over others.

Between the frightfulness of running needles into the flesh and the high improbability of any benefit derived from such a practice, a hundred and seventeen years elapsed before any European practitioner made trial of it. Dujardin in his *Histoire de la Chirurgie*, and Vicq-d'Azyr in the *Encyclopedic Méthodique*, mentioned it above a century after Ten Rhyne had published, but only to congratulate the world that the statements of Ten Rhyne and Kaempfer had not induced any one to practise it, and the first European trials were made by Dr. Berlioz[1] of Paris in 1810. Its power proved so extraordinary that he employed it very extensively, and numerous French practitioners imitated his example with the same results. A body of similar English testimony followed, and acupuncture affords a striking instance of a good remedy discovered from groundless hypothesis, and condemned without a single trial for above a century.

The diseases in which the power of acupuncture is well established are pain and spasm not dependent upon inflammation or organic disease. In rheumatism of the nerves, rheumatic neuralgia, — as distinguished from that chronic form which is generally limited to a small extent of nerve, lasts a great length of time, and is independent of cold, — the invariable causes of rheumatism; in rheumatism of the fleshy parts; in simple pain of any spot; and in spasmodic and convulsive pain of various parts, whether local or migratory, its utility is very great, provided inflammation be not the cause. Of 129 rheumatic cases treated by Dr. Jules Cloquet, about 85 yielded to

---

[1] Berlioz, *Mémoires sur les Maladies Chroniques, les Évacuations sanguines, et l'Acupuncture*. Paris, 1816.

acupuncture. Of 34 published by others, 28 were cured. The writer of this article employed it in St. Thomas's Hospital, and published his results in the 14th vol. of the *Medico-Chirurgical Transactions*. Of 42 cases, taken in succession as they stood in the hospital-books, 30 were found to have been cured: and the remaining 12 had clearly not been adapted for the remedy, as either heat of the affected parts had existed or heat had aggravated the pain. Experience has fully confirmed the fact, that, if rheumatism be at all inflammatory, — be accompanied by heat, or aggravated by a high degree of heat, even though a moderate degree do not aggravate the pain, no relief is in general to be expected from acupuncture. The omission of this distinction and of a little trouble to make it with nicety, will be the chief cause of the operation proving unsuccessful in rheumatism.

In some cases of inflammation and organic disease, however, when pain has been felt apparently disproportionate to those affections, acupuncture is said to have afforded relief.

The pain both of rheumatism and of some nervous affections has occasionally shifted its seat on the application of the needles, and yielded to their repetition in its new situation. Sometimes it required longer chasing from part to part before it vanished.

The cures of ophthalmia, blindness, asthma, diplopia, and hooping cough, by this remedy, must be regarded as lucky occurrences.

The needles employed in Europe are of steel; long and fine; and furnished with either a knob of sealing-wax at their head, or, what is more convenient, a little handle of ivory or wood, screwing into a sheath for the needle. {They are usually from two to four inches long, the length being adapted to the depth it may be desired to make them penetrate. If steel needles be selected, they should be heated to redness and allowed to cool slowly, in order that they may be less brittle. At the blunt extremity of the needle a head of lead or sealing-wax is attached to prevent it from being forced entirely into the body. This is the simplest method of acupuncture, and it is as effectual as any other. If a needle-holder, or *porte-aiguille*, be used, that recommended by Professor F. Bache is as good as any that has been invented. The needle with its *porte-aiguille* consists of a handle, with a steel socket, to receive the end of the needle, which may be fixed securely after having been inserted by the pressure of a small lateral screw. By this contrivance the operator can at pleasure fix in the handle a needle of the length he may desire, and, after inserting it, he is enabled to detach the handle by releasing the screw. After all, however, needles prepared in the simple manner mentioned above are adequate to every useful purpose.} They are best introduced by a slight pressure, and a semi-rotatory motion, between the thumb and fore-finger; and withdrawn with the same motion. The pain is next to nothing, and often absolutely nothing.

The operation may be performed in muscular, aponeurotic, and tendinous regions; and the needle introduced to the depth of from 1/4 of an inch to 2 inches, according to the thickness of the muscles. We should not advise it to be passed into viscera, articulations, or blood-vessels. In general no fluid escapes when the needle is removed; but now and then a small drop of blood follows; and once when the needle had been introduced into the pectoral muscle, I knew blood to spirt forth, but it was immediately restrained by gentle pressure, — an occurrence in every respect similar to what once happened in the practice of M. Bretonneau.[1]

The period during which the needle remains in the part is a matter of great importance. The pain may indeed cease instantaneously: but more frequently does not till the

---

[1] M. Bretonneau says, that he had passed needles into the cerebrum, cerebellumi, heart, lungs, and stomach, of sucking puppies, through and through, and in all directions, with no sign of pain nor particular ill effect; unless when too large a needle was thrust into the heart, and in one instance of this, a little extravasation took place into the pericardium. So far from fearing to acupuncture the heart, Dr. Carraco would have ns do so in the worst cases of asphyxia. He declares that, in the presence of several persons, he kept several kittens under cold water till they were apparently dead,— stiff, motionless, frothing at the mouth, without pulsation of the heart, — and regularly sunk to the bottom every time they were thrown into the water again; that he passed a needle into the heart; that soon the needle began to be gently agitated, then rapidly so, and one voluntary motion after another gradually recommenced, till life was fully re-established; and that the animals did as well afterwards as if nothing had happened.
Death, however, by acupuncture of the brain or spinal marrow, as a secret mode of infanticide, is notorious in works on State-Medicine. "Guy Patin relates that a midwife was executed at Paris who had murdered several infants, at the moment their head presented at the os uteri, by passing a long and very fine needle into the brain through the temples, the fontanelle, or the nape of the neck, or into the heart and its large vessels. Alberti and Brendel quote similar examples. In the *Causes Célèbres* we read the horrible story of a woman who, towards the middle of the last century, made it her business to murder all the new-born infants that fell into her hands by acupuncture, practised at the beginning of the vertebral column, or in the brain, with the sole intention, she told the judges, *of peopling heaven more and more*" [de peupler de plus en plus le ciel] — Fodéré, *Traité de Médecine légale*, 1813, 4:492ff. [See also Belloc, *Cours de médecine légale*, 1819, 96, mentioning the case of needle infanticide.]

needle has remained some time: and my own experience accords with that of others, — that one needle, remaining an hour or more, is far more efficacious than several speedily withdrawn. I usually allow them to remain one or two hours; and have known them remain twenty-four hours, without any ill effect. I have usually found the operation requisite a second time, and in one case, lumbago did not yield till the ninth repetition.

The modus operandi of acupuncture is unknown. It is neither fear nor confidence; since those who care nothing about being acupunctured, and those who laugh at their medical attendant for proposing such a remedy, derive the same benefit, if their case is suitable, as those who are alarmed and those who submit to it with faith. Neither is it counter-irritation; since the same benefit is experienced when not the least pain is occasioned, as when pain is felt. Galvanism, likewise, fails to explain it; because, although the needle frequently becomes oxidated and affords galvanic phenomena while in the body, these phenomena bear no proportion to the benefit, equally take place when acupuncture is practised upon a healthy person, and do not take place when needles of gold or silver are employed, which, however, are equally efficacious with a needle of steel.

{We can scarcely conceive the effects to be anything more than a new nervous impression produced by the needle in the parts which it penetrates.}

Acupuncture has been successfully employed to remove the fluid of oedema and anasarca. In these cases, the needle does not require to be passed deeply; its point has merely to go through the cutis. As soon as this is done and the needle withdrawn, a small bead of water appears at the puncture, which augments till the fluid runs down; and the

oozing will continue for a longer or a shorter time, —
generally for some hours, occasionally for a few days, and
even after death, should that event take place. Any number
of punctures may be made. Although the puncture is so
minute, it is, in such cases, not devoid of danger, any more
than scarification, if practised below the knee. The writer
has frequently had recourse to it with great advantage in
oedema of the scrotum and penis, frequently along the
trunk, and the whole length of the superior extremity, and
on the posterior part of the thigh, and never saw or heard
of the least inconvenience. But several cases have been
related to him, in which sloughing, and in some of which
fatal sloughing, resulted from its performance below the
knee, even though the needle hail been passed merely
through the cutis. Before these cases came to his
knowledge, he had acupunctured the leg, and even the
foot, in dropsy, and never but once saw any inconvenience,
and that was merely a suppuration at each puncture. It
should evidently, however, never be performed below the
knee except when absolutely necessary, a circumstance that
hardly can happen, except in oedema not extending higher
than the knee: and when we reflect that acupuncture
removes an effect only, leaving the cause of the effusion
untouched, and that a large number of effusions are the
result of an inflammatory state, or of sanguineous
congestion, and that, while lessening or removing these by
bleeding, general or local, or purging, we are employing
means which have also a direct tendency to excite
absorption; and when we reflect upon the powers of
diuretics when those measures have previously been
properly employed, we shall perceive that the cases of
dropsical effusion in which acupuncture is required, are
comparatively few.

{Dr. Dunglison has used acupuncture with advantage to drain off the fluid from the cellular membrane in anasarca. In such cases, larger needles are advisable. Some prefer them to be of the size of an ordinary glover's needle, and of a triangular shape, — a puncture of this kind being less likely to close. It has likewise been advised in ascites, in hydrocele, and in every form of encysted dropsy. By M. Velpeau it has been proposed to treat aneurism by acupuncture. In performing some experiments on animals, he found that arteries, punctured by the needle, became the seat of a coagulum, and were ultimately obliterated. In 1830, he read a paper before the Academic Royale des Sciences of Paris, proposing the operation in the cases in question.[1] He found, in his experiments, that whenever the needle remained three days in the flesh, the transfixed artery was completely obliterated.

M. Bonnet treated eleven cases of varicose veins by introducing pins through their cavities, and allowing them to remain there some time. Nine of these cases were cured. The same treatment was applied to hernial sacs. He passed three or four pins through the hernial envelopes close to the inguinal ring, and, in order that they might exert a certain degree of compression, as well as irritation, on the sac, he twisted the points and heads upwards, so as to give them a circular direction. The inflammation usually

---

[1] Velpeau, 'Memoir on the acupuncturation of arteries in the treatment of aneurism', *London medical gazette*, 1830-31, 7:497-499 ; 'Mémoire sur la piqûre ou l'acupuncture des artères dans le traitement des anévrismes', *Gazette médicale de Paris*, 1831, 2:1. 'On the acupuncturation of arteries in the treatment of aneurism', *Amer. Journal of the Medical Sciences*, 1831, 8:510-512. See also a review made by Buet, 'De quelque procédés récemment imaginés pour remplacer la ligature dans les cas d'anévirsme' in *Journal complémentaire des sciences médicales*, 1831, 54-59.

commenced on the third or fourth day, and the pins were removed a few days afterwards. M. Bonnet treated four cases of inguinal hernia in this manner. In two, the hernia was small, and three weeks sufficed for the cure: the third was more troublesome.[1]

Acupuncture has likewise been employed to remove a ganglion of considerable size on the extensor tendons of the foot. After the needle was inserted, pressure was applied, and within a week the tumour had entirely disappeared.[2] When acupuncturation is conjoined with galvanism or electricity, it *constitutes galvano-puncture* and *electro-puncture* (q. v.). See, on the whole subject of acupuncture, Dunglison's *New Remedies*, 4th edit., p. 53. Philad. 1843.}

John Elliotson.

---

[1] *Bulletin général de thérapeutique*, 'Cure of varicose veins and hernia by acupuncturation', *American medical intelligencer*, 1837-38, 1:317; and *Archives Générales de Médecine*, Mai, 1839.
[2] Mr. Vowell, 'Acupuncture of ganglions', *Lancet*, Aug. 25, 1838, p. 770.

5. Acupuncture – Dunglison, 1839

New remedies: The method of preparing and administering them; their effects on the healthy and diseased economy, &c., 1839, 23-30, 163-165, 400-403, 403-406.

Robley Dunglison. Acupuncture

Acupunctura

Synonymes. Acupuncture; Acupuncturatioo.

German. Die Akupunktur; der Nadelstich.

Although acupuncturation is really an ancient therapeutical agent, attention to it has been so much revived of late years, and its use has been so largely extended, that it may be looked upon as constituting one of the novelties of therapeutics.

It consists in the introduction of needles into different parts of the body with the view of removing or mitigating disease; and appears to have been entirely unknown to the Grecian, Roman, and Arabian physicians.[1] From the most ancient times, however, it has been in use with the Chinese and Japanese, by whom it was regarded as one of the most important of remedial agencies. By these people it was systematically taught on appropriate phantoms or mannekins, called *Tsoe-Bosi*, and the practice of the operation was permitted to those only who were able to

---

[1] V. A. Riecke, Die neuern Arzneimittel u. s. w. S. 12, Stuttgart, 1837.

pass a rigid examination thereon. In Europe, it was first known about 156 years ago, from the writings of a Dutch surgeon, Ten-Rhyne, who wrote in 1683;[1] and attention was subsequently drawn to it by Kämpfer;[2] but after this it was almost forgotten, until Berlioz, in 1816, drew attention to its employment. His example was soon followed by Beclard,[3] Bretonneau,[4] Haime,[5] Demours,[6] Sarlandière,[7] Pelletan, Ségalas, Dantu, Velpeau, Meyranx,[8] Dance, in France; by Churchill, Scott, Elliotson,[9] and others, in England; by Friedrich,[10] Bernstein,[11] L. W. Sachs, Heyfelder, Michaelis,[12] Grafe,[13] and others, in Germany; by Carraro,[14] Bergamaschi,[15] Bellini, and others, in Italy; and by Ewing,[16] E. J. Coxe,[17] Bache,[18] and others, in this country.

---

[1] Mantissa schematica de acupunctura ad dissert. de arthritide. London, 1683.

[2] Amaonitat. Exotic. politico-physico-medic. p. 583. Lemgov. 1712; and his History of Japan, vol. ii. Appendix, sect. 4, p. 34.

[3] Mém. de la Société Médic. d'Emulation, viii. 575.

[4] Journal Universel des Sciences Med. xiii. 35. Paris, 1817.

[5] Journal Génér. de Médec. tom. xiii. and Journal Univers. des Sciences Médic., tom. xiii. 1819.

[6] Ibid. tom. xv.

[7] Mém. sur l'Electropuncture. Paris, 1825.

[8] Archives Générales de Méd. tom. vii. Paris, 1825.

[9] Med. Chir. Trans, xiii. 467. Lond. 1827; and art. Acupuncture, in Cyclop. Pract. Med. Lond. 1832.

[10] Translation of Churchill's work in German, p. 40.

[11] Hufeland's Journal, lxvii. Berlin, 1828.

[12] Gräfe und Walther's Journal, B. v. St. 3. S. 552.

[13] Grafe, in art. Acupunctur, in Encyc. WOrterb. der medicinisch. Wissenschaft. B. i. S. 312. Berlin, 1828.

[14] Annali Universal d'Omodei, 1825.

[15] Ibid. 1826.

[16] North Amer. Med. and Surg. Journal, ii. 77. Philad. 1826.

[17] Ibid. ii. 276. Philad. 1826.

[18] Ibid. i. 311. Philad. 1826; and art. Acupuncture, in Cyclop, of Pract. Med. i. 200. Philad. 1833.

M. Jules Cloquet had much to do in reviving its employment in his own country and elsewhere, by his treatise on the subject published at Paris, in 1826, where it was for a long period a fashionable article in the hospitals; so much so, it is affirmed, that attempts were even made to heal a fractured bone by it without the application of any appropriate apparatus! and at one time, it is said, the patients in one of the hospitals actually revolted against the *piqueurs médecins!*[1]

MODE OF ADMINISTRATION.

In the operation of acupuncture, needles are employed, which are very fine, well-polished and sharp pointed. They are usually from two to four inches long, the length being adapted to the depth it may be desired to make them penetrate. If steel needles are selected, they are heated to redness, and allowed to cool slowly, in order that they may be less brittle. At the blunt extremity of the needle, a head of lead or sealingwax is attached to prevent it from being forced entirely into the body. This is the simplest method of acupuncturation, and it is as effectual as any other. By various acupuncturists, needle-holders or handles of ivory have been devised, to some of which the needle is permanently attached. Perhaps the porte-aiguille or needle-holder recommended by Dr. F. Bache,[2] of this city is as good as any that has been invented. The needle, with its porte-aiguille, consists of a handle with a steel socket to receive the end of the needle, which may be fixed securely, after

---

[1] Riecke, Op. cit. S. 13.
[2] Cyclop, p. 202.

having been inserted, by the pressure of a small lateral screw. By this construction, the operator can at pleasure fix in the handle a needle of such length as he may desire, and after inserting it he is enabled to detach the handle by relaxing the screw. After all, however, needles prepared in the simple manner mentioned above, are adequate to every useful purpose.

Besides the common steel needles, those of gold, silver, and platina have been used, but it does not appear that one metal is preferable to another.

To introduce the needle, the skin is stretched, and the needle inserted by a movement of rotation performed in opposite ' directions, aided by gentle pressure. As a rule, the seat of pain will indicate the place where the needle should be introduced; but where the feelings of the patient do not indicate the spot, it must be suggested by our knowledge of anatomy and physiology. From the experiments of Béclard, Brétonneau, Ségalas, Dantu, Velpeau, and others, it would appear, that perforation of arteries, nerves, and even of important viscera with very fine needles has not been followed by any injurious results; yet, at times, accidents have been- produced thereby; and, therefore, it may be laid down as a rule, that the greater nerves, and arteries of a certain size, should be avoided. Prudence would likewise suggest, that important viscera, as the heart, stomach, intestines, etc. should not be penetrated.

The number of needles to be used varies according to the extent of the affected parts. In the opinion of many experienced physicians, we ought not to be afraid of the number, but rather insert too many than too few, and not at too great a distance from each other.

The length of time, during which the needle should be suffered to remain in the part, differs; no fixed rule can be laid down. Some suffer them to remain for an hour and a half or two hours; at times, a period of five minutes is sufficient. In other cases, they have been kept in for two or three days. It appears to be by no means settled what medicinal influence is exerted by their longer or shorter continuance in the parts.

EFFECTS ON THE ECONOMY.

We have already alluded to the impunity with which, in the generality of cases, acupuncturation may be practised even on important organs.

As respects the nerves, Cloquet has seldom or never seen the puncture of them give rise to so much pain as to render it necessary to withdraw the needles; the pain was generally trifling and speedily passed away. He inserted needles into the brain and spinal marrow, and into the crural nerve of a cat, without any evidence of severe suffering or of change of function. Similar experiments were made by E. Gräfe with the same results.[1]

Nor was inconvenience found by Delaunay, Beclard and Cloquet to be sustained in puncturing the arteries and veins. A few drops of blood perhaps issued, and the flow was readily stopped by pressure with the finger. The slight ecchymosis, which, at times, supervened, disappeared rapidly of itself. In Gräfe's experiments, he never found much bleeding to ensue, although he properly esteemed it

---

[1] Art. Acupunctur, in Encyc. Wörterb. u. s. w. S. 317. Berlin. 1828.

advisable to keep clear of the nerves and blood vessels, in order to avoid any unnecessary pain or mischief. As regards the fascia? and periosteum, Gräfe found, that the insertion of needles into them was always very painful, and he re commends, therefore, that the operation should be performed with care on those parts. Should, however, the needles be introduced, and much pain be experienced, it rapidly ceases when they are withdrawn.

Lastly—MM. Haime, Brétonnean, Velpeau, and Meyranx, instituted several experiments on dogs by passing needles into the brain, heart, lungs, stomach, &c. and little or no inconvenience, as we have remarked above, was experienced, provided the needles were extremely fine. Cloquet passed his needles so deeply into the chest of an animal as to leave no doubt that they had penetrated the lungs, and he subsequently pierced the liver, stomach, and testicles without the supervention of any unpleasant results.

The pain occasioned by acupuncturation is generally easily tolerated, but at times it is so violent, that the patients cry out; the violence, however, usually passes away either when the needle is drawn out or forced in deeper. It would seem that the operation is, as a general rule, most successful when it occasions the least pain. Cloquet asserts, that a kind of electric shock is sometimes experienced in the surrounding parts at the moment of the introduction of the needle; in other cases, a tremulous motion is observable in the fibres of the muscles penetrated. Almost always, sometime after the entrance of the needles, a more or less regular areola or halo of a red colour, and without tumefaction, is perceptible around the needles, which soon disappears after they are withdrawn; but when they are suffered to remain long in the part it may persist for hours.

When the operation is productive of benefit, relief is speedily experienced.

The extraction of the steel needles is ordinarily accompanied by more pain than their insertion, especially if they have penetrated deeply, and been retained in the flesh for a long time. The difficulty is owing to their having become oxidised, and consequently rough on the surface. In withdrawing them, it is advisable to give them a movement of rotation, and at the same time to press upon the skin surrounding them with the thumb and index finger.

In the hospitals of St. Louis, La Pitié, and Hôtel Dieu, of Paris, acupuncturation was practised some thousands of times, and in every case, according to Guersent, without the occurrence of anything unpleasant. Pelletan, however, affirms, that he saw it on four occasions followed by slight faintness at the hospital St Louis, but none of the cases assumed the characters of full syncope. Gautier de Claubry has frequently seen faintness, febrile movements, spasm, and insupportable pain produced by it, and Heyfelder saw it followed by convulsions and fainting. Béclard has related a case where the needle penetrated to the bone, and occasioned intense pain. The patient remained a long time faint, and afterwards violent delirium ensued, which gradually ceased in the course of the day, and was followed by great debility. Subsequently, an abscess formed in the part in which the operation was practised.

As to the *modus operandi* of acupuncturation, we cannot conceive its effects to be anything more than the new nervous impression, produced by the needle in the parts which it penetrates. The needles having been found oxidised, especially at the point, it has been supposed by some that the oxidation is connected with the remedial agency, and it has been even affirmed, that in some diseases they oxidise more readily than in others.[1] It is a sufficient reply to this view, that beneficial results are obtained from the use of needles made of metals that do not become oxidised, and that the steel needles oxidise in the sound as well as in the diseased body, and even in parts that have been removed from the body, and placed in warm water; for in the cold dead body, it is affirmed, the phenomenon is not observed. Cloquet and Pelletan think, from their experiments, that the effects of acupuncturation are a consequence of the development of the nervous fluid—which they liken to the galvanic—around the needles; a view which is denied by Pouillet and Béclard, but adopted in a modified form by Dr. Bache,[2] who throws out the conjecture, "that in many cases of local pain this accumulation of the nervous (electrical) fluid depends upon the altered state of the various fasciae or condensed sheets of tissue, giving them the power, to a certain extent, of insulating the parts which they serve to embrace." The explanation is ingenious, but we do not think it necessary, if adequate, to explain the phenomena. We have no doubt, that the effects are owing to a concentration of the nervous power towards the part transfixed by the needle, so that a derivation of the nervous influx is induced towards the seat of pain or towards the nerves particularly concerned in the production of the pain; but further than this we know not.

There is one phenomenon, by the way, which is dependent on the oxidation of the needle. When the free extremity of an inserted needle is connected with the ground by means of a conducting substance, or is put in connection with a soft part of the patient's body, it becomes the seat of a galvanic current, which is exhibited by the multiplier of

---

[1] Gräfe loc. cit. S. 319.
[2] Op. citat. 305.

Schweiger. That this phenomenon is dependent upon the oxidation of the metal is proved by the circumstance, that it does not take place when an unoxidisable metal is employed.[1]

Acupuncturation has been used by Berlioz[2] in gouty and rheumatic cases; by Haime in rheumatic, spasmodic, and convulsive affections, and by Demours in amaurosis and ophthalmia, the needles being inserted in the temples; Finch advised it in an asarca practised on the feet; he also discharged, in this way, the fluid of ascites.[3] Pipelet[4] employed it advantageously in a violent convulsive affection. The needles did not remove or markedly diminish the symptoms, but they postponed their recurrence. Michaelis[5] cured a case of rheumatism by it, but he did not find it so serviceable in oedema of the feet, as the fluid would not flow readily through the minute apertures. Friederich proposed, that in cases of asphyxia, when every other remedy had been employed unsuccessfully, the cavities of the heart should be penetrated by a needle to excite its contraction, and this plan was subsequently ad vised by Carraro,[6] who found, from his experiments on cats, that they could in this way be resuscitated after drowning, when every manifestation of vitality had ceased. His experiments, however, when repeated by Dr. E. J. Coxe,[7] of Philadelphia, were not found to succeed. J. Cloquet obtained the happiest results from acupuncturation in neuralgia, rheumatism, muscular contractions, spasms,

---

[1] Riecke, S. 16.
[2] Op. citat. Paris, 1816.
[3] Lond. Med. Repos. Mar. 1823.
[4] Journal Complem. du Dict. des Sciences Médic. t. xvi. 1823.
[5] Gräfe and Walther's Journal, B. v. St. 3.
[6] Annal. univ. di Medicin. 1825.
[7] North Amer. Med. and Surg. Journal, ii. 292.

pleurodyne, cephalalgia, ophthalmia, toothache, epilepsy, gout, gastrodynia, contusions, lumbago, periodical amaurosis, diplopia, paralysis, etc.

It is in rheumatic affections that its success has been most marked. Dr. Elliotson[1] cured 30 out of 42 cases by it in St. Thomas's hospital. In sciatica its efficacy has been evident.[2]

By Velpeau it has been proposed to cure aneurism by acupuncturation. In performing some experiments on animals he found, that arteries punctured by the needle became the seat of a coagulum, and were ultimately obliterated. In 1830, he read a paper before the Académie des Sciences, of Paris, proposing the operation in the cases in question.[3] He found in his experiments, that whenever the needle remained three days in the flesh, the transfixed artery was completely obliterated.

M. Bonnet, Chirurgien-en-chef to the Hôtel Dieu at Lyons,[4] has affirmed, that he treated eleven cases of varicose veins by introducing pins through their cavities, and allowing them to remain there some time. Nine of these cases were cured. The same treatment was applied to herniary sacs. He passed three or four pins through the herniary envelopes close to the inguinal ring, and in order that they might exert a certain degree of compression, as well as of irritation, on the sac, he twisted the points and heads upwards so as to give them a circular direction. The inflammation and pain

---

[1] Art. Acupuncture, Cyclop. pract. Med.; Lond. 1832.
[2] Renton, in Edinb. Med. and Surg. Journ. for 1830, xxxiv, 100, and Dr. Graves in Lond. Med. Gaz. July, 1831, and Lond. Med. and Surg. Journal, April, 1833.
[3] Lond. Med. Gaz. and Amer. Journal Med. Sciences, Aug. 1831, p. 510.
[4] Bulletin Generate de Thérapeutique, and Dunglison's American Intelligencer, for Dec. 1, 1837, p. 317.

usually commenced on the third or fourth day after the operation, and the pins were removed a few days afterwards. M. Bonnet had treated four cases of inguinal hernia by acupuncturation. In two, the hernia was small, and three weeks sufficed for the cure: the third was more troublesome.

Caution is of course requisite not to injure the spermatic cord.

Of late, acupuncturation has been revived[1] in the treatment of hydrocele by Mr. Lewis, Mr. King,[2] and others.

It consists in carrying a common sewing needle—of the size used for sewing a button to a shirt— through the skin, the dartos and cremaster, into the bag containing the fluid, so that a drop of the fluid follows the instrument as it is withdrawn. It is executed in nearly the same manner as the ordinary method of tapping with a trocar, except that the needle, which should be oiled, cannot be plunged in so easily as that instrument. Mr. King suggests that the needle should be fixed in a handle, by which means it can be made to enter with comparative facility. After the operation, a compress, moistened with a discutient lotion, may be kept on the scrotum, and the patient may walk about or remain at rest, as may best suit him.

The phenomena which present themselves in a few hours are as follows:—the swelling begins to be less circumscribed, and to lose its tenseness, and the cellular tissue of the scrotum becomes gradually more and more infiltrated with the fluid, which before dis tended the tunica vaginalis, and which, in the space of from twenty-four to

---

[1] Mr. Travers, in Lond. Med. Gazette, Feb. 1837, p. 737. Mr. Lewis, Ibid. Feb. 1837, p. 788. Mr. Robert Keate, Ibid. p. 789.
[2] British Annals of Medicine, No. 1, p. 13.

forty-eight hours, will, according to Mr. King, have entirely changed place. In Ave or six days, the infiltration disappears, and the patient is cured.

Mr. Lewis first introduced the method as a palliative cure, but he has seen cases where a radical cure was effected by it.[1] He considers the principle of puncturing with a fine pointed needle not only applicable to promote the absorption of the fluid in hydrocele, but in every case of encysted dropsy.[2]

We have already referred to the use of acupuncturation in anasarca. We have used it advantage ouslyin these cases to drain off the fluid from the cellular membrane; in such cases larger needles are needed; some prefer them to be of the size of an ordinary glover's needle, and of a triangular shape: a puncture of this kind being less likely to close.[3]

In the mass of cases, it need scarcely be said, this course can act merely in a palliative manner, the cause of the dropsical accumulation still persisting. Still, as Dr. Graves has remarked, under favourable circumstances and in a good constitution, the simple operation of evacuating the fluid by punctures made through the skin, has been, of itself, sufficient to effect a cure. In a lady, under his care, a general anasarca came on after fever, and resisted every form of treatment he could devise. When he had made many fruitless attempts to produce absorption by means of internal remedies, another practitioner was called in, who practised acupuncturation of the lower extremities, and succeeded completely.

---

[1] Dr. Davidson, in Edin. Med. and Surg. Journal for Jan. 1838.
[2] Lancet, May 7, 1836, and Jan. 14, 1837.
[3] Dr. Graves, Lond. Med. Gazette, Oct. 1838. See, also, Mr. King, Ibid. Oct. 7, 1837, and Nov. 25, 1837.

Lastly, Mr. Vowell[1] has published a case in which acupuncturation was successfully employed for the removal of a ganglion. A young lady under his care had been affected with a ganglion of a considerable size on the extensor tendons of the foot, which produced not only disfiguration, but some uneasiness. Mr. Vowell applied blisters, and afterwards the iodine ointment and pressure, for above a month, without benefit. He then inserted the tambour porte-aiguille of his patient. Pressure was applied, and within a week the tumour had completely disappeared.

When acupuncturation is conjoined with electricity or galvanism, it constitutes electro-puncture, and galvano-puncture.

---

[1] Lancet, Aug. 25, 1838, p. 770.

Electropunctura

Synonyme.—Electropuncture.

This consists in a union of acupuncturation with electricity.

The idea of the conjunction appears to have originated with Berlioz; but Sarlandiere was, doubtless, the first who put it in practice, although J. Cloquet has contested the priority with him—a matter, by the way, as in all such cases, of extremely small moment. The operation consists in employing acupuncturation in the usual way, either with a single needle, or with two or more; and making a communication between them and the prime conductor of an electrical machine; or they may be made to form part of the circuit in the discharge of a Leyden jar. In this way, the electrical influence may be graduated from the simple aura to a full shock. Sarlandiere appears to have employed electropuncture with great success, but he restricts its use to rheumatic or neuralgic pains, uncomplicated with organic mischief or inflammation: when such complications exist, he advises bloodletting and general antiphlogistics to be premised.[1] Guersent thinks it better, in all these cases, to use simple acupuncturation, and only to employ electropuncture, when the first proves to be inadequate, as in paralysis, and in tremors produced by the immoderate use of mercury; — in all cases, indeed, in which the malady depends on a diminution of the nervous energy. A case of success from its use, in paralysis of the right arm, in which voluntary motion and sensibility were destroyed, has been

---

[1] E. Gräfe, Art. Electropunctura, in Encyclopäd. Wörteib. der medicinisch.Wissensch. x. 550. Berlin, 1834.

recently published.[1] The patient was, in the first instance, subjected to the use of blisters and moxas along the course of the radial nerve, from which he obtained some advantage. The remedy which succeeded best, however, was electropuncture along the nerves from the shoulder to the hand. At first, the punctures were but little felt, but afterwards they were very painful. The sensibility, mobility, and strength of the fingers and hand gradually returned; and, three months after his admission, the patient left the hospital completely cured.

Magendie affirms, that he has treated many cases of incomplete amaurosis with great success by this agency. He employed it, however, in the form of what has been more properly termed galvanopuncture; by fixing a needle in the frontal nerve, and another in the superior maxillary, and making these communicate respectively with the poles of a galvanic pile of twelve pairs of plates, each six inches square. Whenever the contact was made, the patient experienced a painful commotion in the course of the nerves, and at the bottom of the orbit; the light became better appreciated, and the pupil contracted.

We have frequently used both electropuncture and galvanopuncture in rheumatic and neuralgic affections; but do not think the advantages were more marked than those of simple acupuncture, whilst the suffering from the operation was certainly greater.

In cases of asphyxia, galvanopuncture has been proposed to arouse the dormant energies. The effect of electricity, in the different forms in which it is adopted in medicine, on the functions of sensibility and muscular contraction, could not

---

[1] La Lancette Françoise, Dec. 20, 1836; and American Med. Intelligencer, Oct. 16, 1837, p. 265.

fail to suggest it early to observers as a fit agent for this purpose; but it is rarely at hand, and, therefore, seldom available. J. P. Frank, Thillaye,[1] and others have highly recommended it; — the latter gentleman, on the strength of numerous experiments on animals. As the object, in these cases, is to arouse the respiratory muscles to action, the electric shock may be passed through the shoulders or the chest in any direction. Neither common nor galvanic electricity is possessed of any power to restore the action of the involuntary muscles. We have frequently attempted to re-excite the action of the heart, intestines, fibres of the uterus, &c. soon after the cessation of respiration and circulation, by means of the galvanic stimulus, but without the slightest success, although the voluntary muscles responded to it energetically. Besides, were the action of the heart re-excited by it, this could be but momentary. A necessary stimulating agency to that viscus is distension by the proper fluid, and unless the respiratory movements were restored, and conversion of venous to arterial blood effected, so that the latter could reach the left heart, the action of that organ could not be maintained. Every attempt, therefore, is properly made to restore the action of the respiratory muscles, so that haematosis may be accomplished.[2]

M. Leroy d'Etioles[3] has suggested galvanopuncture in a manner which, at the first aspect, appears most formidable; but which is really less so than it seems to be, in

---

[1] Archives Générales de Médecine, xii.

[2] Art. Asphyxia, by the author, in the American Cyclopaedia of Practical Medicine, part x. p. 486, Sept. 18-6.

[3] Magendie's Journal de Physiologie, tom. vii. 1827; tom. viii. and tom. ix; also Recherches Expérimentales sur l'Asphyxie, Paris, 1829; and Bourgeois, Observations sur la possibilité du retour & la vie dans plusieurs cas d'Asphyxie. Paris, 1829.

consequence of the impunity with which fine needles can be made to penetrate, as we have seen,[1] even the most important organs. He introduced an acupuncture needle on each side between the eighth and ninth rib, until it reached the fibres of the diaphragm. He then established a galvanic current between the needles by means of a pile of twenty-five or thirty pairs of plates, an inch in diameter. The diaphragm immediately contracted, and an inspiration was accomplished. He then interrupted the circle, when the diaphragm, urged by the weight of the abdominal viscera, and aided by gentle pressure made on the abdomen by the hand, returned to its former position, and an expiration was effected. In this way, the two respiratory acts were made to succeed each other, and regular respiration was reinduced. A continuous current was likewise employed in some cases, but the respiratory movements were irregular, and nothing like natural respiration resulted.

Leroy tried his method on animals asphyxied by submersion, and when they had not been under water more than five minutes, they were often restored.

These experiments were witnessed by Magendie.[2] On different occasions, M. Leroy asphyxied animals of the same kind, and apparently of the same strength, and whilst those that were left to themselves perished, those that were treated by galvanism recovered.

As an aid, therefore, to pulmonary insufflation, and an important one, galvanism, combined or not with acupuncturation, might be advantageously employed in asphyxia, but as has been already remarked, it can rarely be available. Certainly no time should be lost in adopting the

---

[1] See Art. Acupuncture.
[2] Journal de Physiologie, ix.

other energetic and indispensable measures that are demanded.[1] It has been recommended, that as only a very small apparatus is necessary, batteries, consisting of a few plates, might be kept wherever there are station-houses for the reception of persons in a state of asphyxia.[2] The suggestion is good; and they might also with propriety form a part of the cabinet of apparatus of the private practitioner; but whilst an assistant is preparing the apparatus for action, the practitioner should be assiduously engaged in applying other means of resuscitation.[3]

---

[1] See Art. Asphyxia, Op. cit. p. 486.

[2] Kay, in Edinb. Med. and Surg. Journ. xxix. and in his work on Asphyxia. Lond. 1834.

[3] See Most, Art. Galvanismus, in Encyklopäd. der gesammten medicin. und chirurgisch. Praxis, u. s. w. 2te Auflage. Leipz. 1836.

Moxa

Synonyme.—Moxiburium.

By the term *moxa*, the Chinese and Japanese designate a cottony substance, which they prepare by beating the dried leaves of the artemisia chinensis, a kind of mugwort. With this down they form a cone, which is placed upon the part intended to be cauterised, and is set fire to at the top.

This mode of exciting counter-irritation has been long practised by the Chinese and Japanese, and by the ruder nations of the old world; but it was not much employed in Great Britain and France until about the commencement of the seventeenth century, when it was introduced through the agency of a physician[1] who had resided in India. It fell again, however, into disuse, until attention was redirected to it, during the last century, by Pouteau[2] and Dujardin, and, at the commencement of this century, by Percy and Laurent,[3] Larrey and others.[4]

---

[1] Ten Rhyne, Medit. de veteri Medicin.; Dissert, de Anthritide, Lugd.Bat. 1672; and Kaempfer's History of Japan, translated by Scheuchzer, vol.ii. append, sect. iv. Lond. 1728.

[2] Mélanges de Chirurgie, p. 49.

[3] Dictionnaire des Sciences Médicales, Art. Moxibustion.

[4] See, for a history of the moxa, the author's translation of Baron Larrey's Memoir on the use of the Moxa. Lond. 1822.

MODE OF PREPARING.

Various agents have been used by different people, in "moxibustion," for so the mode of cauterisation has been termed, which consists in placing some combustible substance on a part of the body, and suffering it to burn down. From the earliest ages, the Nomades employed the fat wool of their flocks, as well as certain spongy substances growing upon oaks,[1] or springing from the hazel;[2] the Indian the pith of the reed,[3] and flax or hemp impregnated with some combustible material;[4] the Persian, the dung of the goat; the Armenian, the agaric of the oak; the Chinese and Japanese, the down of the artemisia; the Thessalian, dried moss;[5] the Egyptians, the Arracanese, and several oriental nations, cotton;[6] the Ostiaks[7] and the Laplanders,[8] the agaric of the birch; and the aborigines of this continent, rotten and dried wood. Hippocrates[9] was in the habit of employing fungi and flax for the same purpose.

In modern times, also, various substances have been used for the fabrication of the moxas. Whatever article is selected, it ought to be a spongy, light, vegetable matter; readily combustible, and so prepared as to burn down

---

[1] Hippoc. lib. de Affect, cap. xxx.
[2] Paulus Aegineta, lib. vi. cap. 49.
[3] Kaempfer, vol. ii. app. sect. iv. p. 36.
[4] Bontius de Medicina Indorum, p. 32.
[5] Percy, in Pyrotechnie Chirurgicale pratique, p. 12.
[6] Prosper Alpini, de Medicina Aegyptiorum, lib. iii. cap. 12.
[7] Voyages de M. Pallas, iv. 68.
[8] Acerbi's Travels through Sweden, Finland, and Lapland, ii. 291, and Linnaeus, in Lachesis Lapponica, translated by Sir James Smith, i. 274.
[9] De Affect, cap. viii.

slowly. In Germany, they use the tinder—*amadou*— which
is known to be an agaric prepared for the purpose; and it is
not uncommonly employed in our hospitals,—a small disc
or cylinder being placed on the part, and set fire to. The
match used by artillerists was recommended by Percy,[1]
after Bontius:[2] it is composed of hemp steeped in a solution
of nitre. He likewise proposed the pith of the sunflower—
helianthus annuus—recommending, that the stalk should
be cut into cylinders of the desired length, the bark being
left on; so that, when ignited, it may burn in the centre and
be held with the hand.[3] This, he calls *moxa de voleurs*.[4] The
moxa, used by Larrey, und very generally employed by
many practitioners, is made by taking n quantity of cotton
wool, pressing it somewhat closely together, and rolling
over it a piece of fine linen, which is fastened at the side by
a few stitches. Larrey advises, that it should have the shape
of a truncated cone—the form usually adopted—and be
about an inch long. Commonly the cylinder is shorter than
this; six or eight lines— as, when above six linos high, the
combustion is not felt—and about four or five lines broad.
The moxas, employed by Dr. Sadler,[5] of St. Petersburg, are
about half an inch in diameter, and three quarters of an
inch in height. They are composed of a nucleus formed of
the pith of the sunflower, wrapped in layers of cotton, of
various thickness, and surrounded with an external
envelope of thin muslin; both of the latter being previously
steeped in a solution of nitre. They are held, while burning,
by means of two long hair pins, the legs of which are slightly

---

[1] Op. cit. p. 77. Paris, 1811.
[2] Op. cit. p. 32. Paris, 1645.
[3] Art. Moxibustion, in Dict. des Sciences Médicales.
[4] Mérat & De Lens, Dict. de Mat. Méd., Art. Moxa.
[5] Zeitschrift für die gesammte Medicin. B. iii. H. ii. & iii. and British
and Foreign Medical Review, July, 1837, p. 217.

bent, in order to accommodate them to the shape of the moxa; and, when the latter is burned down to the place where it is held by the first hair-pin, it can be held with the other, and retained in its proper position. With this last view, Larrey[1] has a special *porte-moxa*, consisting of a ring to receive the cylinder, with a handle attached to it, and three small supports or knobs of ebony, placed beneath the ring, to prevent the heated metal from acting upon the surface.

Of late years, a plan for raising vesication on the surface has been adopted, which, as Dr. Granville remarks, must be regarded as a kind of moxa.[2] This, he admits, is equally successful with the one he proposes, and which we have already described,[3] in forming a rapid vesication; "but it is, at the same time, so complicated, and attended by such intense pain," that, in practice, he says, it will not bear comparison with the preparations which he recommends. A piece of linen or paper, being cut of the requisite size, is immersed in spirit of wine, or brandy.

It is then laid on the part to be blistered, care being taken that the moisture from the paper or linen does not wet the surrounding surface.

The flame of a lighted taper is applied quickly over the surface, so as to produce a general ignition, which is exceedingly rapid. At the conclusion of this operation, the cuticle is found detached from the true skin beneath.

---

[1] The author's translation of his Essay on the Moxa, p. 5.
[2] Counter-irritation, its Principles and Practice, Amer. Med. Library edit, p. 21 and p. 42. Philad. 1838.
[3] [Dunglison, New Remedies, in American medical library, 1839, 396.]

In the application of the various moxas, or of most of them, their agency can be so graduated as to produce either simple rubefaction, vesication or the formation of an eschar. Where it is desirable to produce the first result only, the cylinder of cotton may be removed when the pain becomes somewhat severe; or the burning material may be held close to the surface, and be moved gradually along it. In this manner, a counter-irritant effect may be exerted along the spine or any extensive surface. Any burning substance—a lighted coal for example—will answer for this purpose. When vesication is needed, it must be kept on longer; and if it be desirable to produce an eschar, the moxa may have to remain on until it is wholly consumed. Larrey,[1] indeed, advises, that the blowpipe should be occasionally employed to hasten the combustion. When the integument has once become disorganised, the slough will be thrown off in due time, leaving an ulcer. Larrey says the sloughing can be prevented by the application of liquid ammonia[2] to the burnt surface, after the moxa has been removed. This will do when the disorganisation is partial; but we know, from experience, that it often fails.

EFFECTS ON THE ECONOMY IN DISEASE.

The moxa—in its different forms—is doubtless a most valuable agent, where rapid counter-irritation is indicated. It resembles, indeed, in its action, the aminoniated counter-irritants of which we have already treated, and is applicable to the same diseases;—the only difference between them—

[1] Op. citat. p. 5.
[2] Ibid. p. 9.

when cauterisation is effected—being, that the agent in the case of the ammoniated lotion is a *potential*, in that of the moxa an *actual*, cauterant.

The moxa must be regarded as one of our most valuable revellents.

Galvanismus

Synonymes.—Galvanism, Electricitas Animalis, E. Galvanica seu Metallica, Irritamentum Metallorum seu Metallicum.

French. — Galvanisme.

The ordinary effects of common and galvanic electricity and of electro-magnetism are so well known, as to require but little comment. They are decidedly excitant; and, like all excitants, when applied to a part of the frame, are counter-irritant or revellent. All have been employed in paralysis—general and local,—amaurosis, deafness and dumbness of recent duration, asthma, rheumatism, neuralgia, &c. The effect, however, which galvanism exerts on the contractility of the muscular fibre, and the great similarity, in its agency, to the nervous influence,[1] has led to its employment more frequently in the various nervous and spasmodic diseases referred to, and in others belonging to the same class. Resting on his views of the absolute identity between the nervous and the galvanic fluids,[2] Dr. Wilson Philip employed it in many diseases, and especially in asthma.

In a paper read by him before the Royal Society of London, in January, 1816, he details some experiments, which he made on rabbits.

The eighth pair or pneumogastric nerves were divided by incisions made in the neck. After the operation, the parsley, which the animals had eaten, remained unchanged in their

---

[1] See the author's Physiology, i. 88, 3d edit. Philad. 1838.
[2] Experimental Inquiry into the Laws of the Vital Functions. Lond. 1817.

stomachs, and after evincing much difficulty of breathing they seemed to die of suffocation. But when, in other animals, whose nerves had been divided, the galvanic agency was transmitted along the nerve, below its section, to a disc of silver, placed closely in contact with the skin of the animal, opposite to its stomach, no difficulty of breathing occurred. The galvanic action being kept up for twenty-six hours, the rabbits were then killed and the parsley was found digested.

The removal of dyspnoea in these cases led Dr. Philip to employ galvanism as a remedy for asthma; and, by transmitting its influence from the nape of the neck to the pit of the stomach, he gave decided relief in every one of twenty-two cases, of which four were in private practice, and eighteen in the Worcester infirmary. The power employed varied from ten to twenty-five pair of plates. Since then, galvanism has been repeatedly used in such cases, and at times with marked relief. Commonly, however, the plates described hereafter, are employed for this purpose. The disease is unquestionably in the majority of cases dependent upon erethism of the pneumogastric nerves; all the phenomena indicate, that there is a spastic constriction of the small bronchial tubes, occasioned by irritation at the extremities or in the course of the nerve.

The new impression made by the galvanic agency, breaks in upon the concentration of nervous action, by exciting other portions of the nervous system, in the same manner as we observe spasms .or ordinary cramp relieved, or paroxysmal diseases warded off, by agents that are capable of suddenly impressing some part of the nervous system.

Not long after these researches of Dr. Philip, galvanism was employed satisfactorily by Mr. Mansford[1] in a congenerous disease—epilepsy—and his plan was afterwards—although tardily—extended to some other paroxysmal disorders. The mode of application, recommended by Mansford, is as follows: A portion of the cuticle, of the size of a sixpence, is removed by means of a small blister on the back of the neck, as close to the root of the hair as possible; and a similar portion is removed from the hollow, beneath, and on the inside of, the knee, as the most convenient place. To the excoriated surface on the neck, a plate of silver, varying—according to the age of the patient—from the size of a sixpence to that of a half crown, is applied, having attached to its back part a handle or shank, and to its lower edge—and parallel with the shank—a small staple, to which the conducting wire is fastened. This wire passes down the back, until it reaches a belt of chamois leather, buttoned round the waist; it then follows the course of the belt to which it is attached, until it arrives opposite the groin of the side on which we desire to employ it; it then passes down the inside of the thigh, and is fastened to the zinc plate in the same manner as to the silver one. The apparatus, contrived in this way, is thus applied. A small piece of sponge, moistened in water, and corresponding in size to the blistered part of the neck, is first placed directly upon it; over this, a large piece of the same size as the metallic plate, also moistened, is laid, and next to this, the plate itself, which is secured in its situation by a strip of adhesive plaster passed through the shank in its back; another above, and another below it. If these be properly placed, and the wire, which passes down the back be allowed sufficient room that it may not drag, the plate will not be moved from

---

[1] Researches into the nature and causes of Epilepsy, &c, Bath, 1819.

its position by any ordinary motion of the body. The zinc plate is fastened in the same manner, but in place of the second layer of sponge, a piece of muscle answering in size to the zinc plate is interposed; that is—a small piece of moistened sponge being first fitted to the exposed surface below the knee, the piece of muscle moistened, or—what we have found equally effectual and less inconvenient—a piece of moistened flannel[1] follows, and on this the plate of zinc.

The apparatus, thus arranged, will continue, according to Mr. Mansford, in gentle and uninterrupted action from twelve to twenty-four hours, according to circumstances. "This last is the longest period that it can be allowed to go unremoved; the sores require cleaning and dressing, and the surface of the zinc becomes covered with a thick oxide, which must be removed to restore its freedom of action: this may be done by scraping or polishing; but it will be better if removed twice a day, both for the greater security of a permanent action, and for the additional comfort of the patient." The adoption of this plan of treatment in cases of tic douloureux, the confidence reposed by Laennec in the use of galvanic plates on the breast and back in angina pectoris and similar neuralgic affections of the chest, and the communications of Drs. Harris and Chapman, brought it into very extensive use, so that ample trial was given to it in this country both in public and private practice. In three cases, it was—to employ the language of Professor Chapman[2]—"triumphantly directed" by Dr. Harris; but it was only found effectual in affections of the face; and in these cases it had to be persevered in for some time before

---

[1] Dr. Chapman says soft buckskin or parchment. American Journal of the Medical Sciences, Aug. 1834, p. 311.
[2] Op. citat. p. 311.

marked benefit was experienced.[1] About the same period, this mode of applying galvanism was recommended by Dr. Miller,[2] of Washington University, Baltimore, and a case of paraplegia and another of general paralysis were adduced by him in which it was found highly efficacious.

There are doubtless—as we have observed—cases in which the excitant and revulsive agency of galvanism may be employed with advantage, but they are not so numerous as was at one time believed. We have used the plates extensively—in neuralgic cases especially—but have not experienced so much success, as to induce us to advise them frequently, under the inconvenience that necessarily accompanies their employment.

They are, indeed, at this time, but little used.

Some years ago, Professor Von Hildenbrand, of Pavia,[3] recommended, in cases of frontal neuralgia, an *anodyne metallic* or *galvanic brush*, which appears to have been as effectual in his hands as the galvanic plates in those of Dr. Harris. It consists of a bundle of metallic wires not thicker than common knitting needles, firmly tied together by wire of the same material, so as to form a cylinder of about four or five inches long, and an inch or three fourths of an inch in diameter. This is applied to the pained part, previously moistened with a solution of common salt; and, according to Von Hildenbrand, it at times produces relief so instantaneous, that it appears to the patients to act like a charm. In his first experiments, he employed brushes constructed of two kinds of metal,—for instance, of silver

---

[1] Dr. Harris, in Amer. Journal of the Medical Sciences, Aug. 1834, p. 384.
[2] Ibid. p. 321.
[3] Edinburgh Medical and Surgical Journal, April, 1833.

and copper wire, copper and zinc wire, or zinc and brass wire, the individual wires being mutually commingled; but he subsequently ascertained, that bundles of wires of one and the same metal produced an effect scarcely less speedy, and that solid metallic bodies acted in a similar manner, but in a much feebler degree. The nature of the metal he thinks occasions no difference.

It is not probable, however, that, in these cases, galvanism is the agency concerned. Like the metallic tractors of Perkins, the effect is probably induced by the new nervous impression made through the excited imagination of the patient.

ANIMAL MAGNETISM— Mesmerism, Neurogamia, Biogamia, Biomagnetismus, Zoomagnetismus, Tellurismus, Exoneurism, as it has been termed—exerts an anodyne influence in probably the same manner. In highly impressible persons, more or less prolonged impressions made upon the senses—as by the operator looking steadfastly in the eyes of the patient; holding her thumbs or hands in his at the same time, or making passes in front of her—will induce an hysteric or hysteroid condition, in which the patient may fall into what is called " magnetic sleep," of a very sound, and at times cataleptic, character: during the existence of this sleep, she may be insensible to certain irritants, and yet extremely alive to others, so that operations—as the extraction of teeth, and even others of a more serious character—may be performed without eliciting the ordinary evidences of feeling. In cases of delirium tremens, accompanied by watchfulness, in which we have the whole nervous system extremely impressible, sleep may be at times induced by the employment of this agency, which has resisted the ordinary anodynes.[1] Lastly. Of late years, it has been proposed to introduce into the rectum, in cases of constipation, a kind of *galvanic suppository*, made of two metals—zinc and copper—and various forms of instruments have been devised by the prolific imaginations of the inventors; those intended for the rectum simply, were doubtless of advantage, at times, by virtue of the excitation they induced in the nerves of the mucous membrane.

Others, formed somewhat like a bassoon—and so arranged as to have one metal in the mouth and the other in the rectum connected together by metal—did not appear to act differently from those of the simpler form.

Both have gone into disuse, and—as we have said elsewhere[2]—if their efficacy on the frame has not been well marked, they have not failed to minister to the pockets of their inventors.

[1] Dr. Vedder, in American Medical Intelligencer, Feb. 1, 1839, p. 331.
[2] General Therapeutics, p. 248, Philad. 1836.

References

References

Acerbi, Giuseppe. *Travels through Sweden, Finland, and Lapland, to the North Cape, in the Years 1798 and 1799.* Vol. 2. London: J. Mawman, 1802. http://archive.org/details/travelsthroughsw02inacer.

Aegineta, Paulus. *The Seven Books of Paulus Aegineta [4-6] : Translated from the Greek : With a Commentary Embracing a Complete View of the Knowledge Possessed by the Greeks, Romans, and Arabians on All Subjects Connected with Medicine and Surgery.* Edited by Francis Adams. Vol. 2. 3 vols. London: Sydenham Society, 1844. http://archive.org/details/sevenbooksofpaul02pauluoft.

Alpini, Prosper, Jakob de Bondt, and Melchior Guilandinus. *Prosperi Alpini ... Medicina Aegyptiorum : accessit huic editioni ejusdem auctoris liber De balsamo, et rhapontico, ut et Jacobi Bontii Medicina Indorum.* Lugduni Batavorum: Apud Gerardum Potvliet, 1745. http://archive.org/details/prosperialpinime00alpi.

"Asphyxia." In *American Cyclopaedia of Practical Medicine*, 486, Sept. 18-6.

Bache. "Acupuncture." In *Cyclopaedia of the Practice of Medicine*, 1:200. Philadelphia, 1833.

Bache, Franklin. "Cases Illustrative of the Remedial Effects of Acupuncturation." *North American Medical and Surgical Journal* i (1826): 311.

Béclard, Adelon. *Dictionnaire de Médecine.* Vol. 1. Paris: Béchet jeune, 1821. http://gallica.bnf.fr/ark:/12148/bpt6k31190w.

———. "Recherches et Expériences Sur Les Blessures Des Artères." *Mémoires de La Société Médicale D'émulation*, no. viii (1816).

Berlioz, Louis-V.-Joseph. *Mémoires sur les maladies chroniques, les évacuations sanguines et l'acupuncture.* Paris: Croullebois, 1816.

https://books.google.com/books?id=aJmnikG4yP4C.

Bonnet. "Cure of Varicose Veins and Hernia by Acupuncturation." *American Medical Intelligencer*, December 1, 1837, 317.

———. "[Phlébite et Infection Purulente (De La Cautérisation Considérée Comme Moyen de Préveniuier et de Guéir La) ??]." *Bulletin Général de Thérapeutique* 24 (1843): 475–77.

Bontius. *De Medicinâ Indorum*. Paris, 1645.

Bourgeois. *Observations Sur La Possibilité Du Retour & La Vie Dans Plusieurs Cas d'Asphyxie [Galvanopuncture]*. Paris, 1829.

Carraro, Antonio. "Saggio Sui Agopunctura Etc [An Essay on Acupuncture]." *Annali Universali Di Medicina*, 1825.

Chapman, N. "Remarks on Tic Douloureux; with Cases." *The American Journal of the Medical Sciences*, August 1834, 311.

Coxe, E. J. "Observations on Asphyxia from Drowning, to Which Is Added a Case of Resuscitation." *The North American Medical and Surgical Journa* 2 (1826): 276–96.

Davidson, William. "Report of Surgical Cases, Treated in the Glasgow Royal Infirmary, during the Years 1836-7." *The Edinburgh Medical and Surgical Journal*, January 1838. https://books.google.com/books?id=yPkaAQAAMAAJ&.

Demours. "Note Sur l'Acupuncture, Lue À La Séance Du 2 Février 1819, Par M. Demours, Membre Résident, Médecin-Oculiste Du Roi, Etc., Etc." *Journal Général de Médecine, ...* 66 (1819): 161–65.

———. "Seconde Notice Sur l'Acupuncture, Ou Introduction de L'aiguille, sans Que La Ventouse Soit Retirée, À Travers La Portion Du Tissu Cutané, Soulevé Par Cet Instrument ; Lue À La Société de Médecine de Paris, Par M. Demours, Membre Résident, Médecin-Oculiste Du Roi, Président Du Cercle Médical de Paris." *Journal Général de Médecine, ...* 66 (1819): 377–83.

Diderot, Denis, Jean Le Rond d'Alembert, Charles Joseph Panckoucke, Henri Agasse, Tessier (Alexandre-Henri

M.), André Thouin, Auguste-Denis Fougeroux de Bondaroy, et al. *Encyclopédie méthodique: ou par ordre de matières - Médecine*. Panckoucke, 1830. https://books.google.com/books?id=tNhTAAAAYAAJ.

Dujardin, François, and Bernard Peyrilhe. *Histoire de la chirurgie depuis son origine jusqu'à nos jours*. Vol. 1. Paris: Imprimerie royale, 1774. http://gallica.bnf.fr/ark:/12148/bpt6k1042753f.

Dunglison, Robley. "Acupuncture." In *New Remedies: The Method of Preparing and Administering Them; Their Effects on the Healthy and Diseased Economy, &c.*, 3:23–30. Dunglison's American Medical Library. Philadelphia: A. Waldie, 1839. https://books.google.com/books?id=EaJQAAAAYAAJ.

———. *General Therapeutics; Or, Principles of Medical Practice: With Tables of the Chief Remedial Agents, and Their Preparations; and of the Different Poisons and Their Antidotes*. Philadelphia: Carey, et al, 1836. https://books.google.com/books?id=970wAAAAYAAJ.

———. *Human Physiology*. Lea and Blanchard, 1841. https://books.google.com/books?id=dDxx2F92BHkC.

———. *New Remedies, Pharmaceutically and Therapeutically Considered*. 4th ed. Philadelphia: Lea, 1843. http://archive.org/details/newremediespharm00dungu oft.

Elliotson. "Acupuncture." In *The Cyclopædia of Practical Medicine: Comprising Treatises on the Nature and Treatment of Diseases, Materia Medica and Therapeutics, Medical Jurisprudence, Etc. Etc*, 1:54–57. Sherwood, Gilbert, and Piper, 1832. https://archive.org/details/cyclopaediaofpra00forb.

Elliotson, John. "Acupuncture in Rheumatism." *Medico-Chirurgical Transactions* 13 (1827): 467–68.

Ewing, J. Hunter. "Case of Neuralgia Cured by Acupuncturation." *North American Medical and Surgical Journal* 3 (1826): 77–78.

Finch. "Case of Anasarca in Which Acupuncturation Was Successfully Employed." *The London Medical Repository* 19 (March 1823): 205.

Fodéré, François-Emmanuel. *Traité de médecine légale et d'hygiène, publique ou de police de santé*. Vol. 4. Paris: Mame, 1813. http://gallica.bnf.fr/ark:/12148/bpt6k76979m.

Gräfe, E. "Acupunctur." In *Encyc. Wörterb. Der Medicinisch.* Berlin, 1828.

———. "Electropunctura." In *Encyc. Wörterb. Der Medicinisch.* Berlin, 1834.

Granville, Augustus Bozzi. *Counter-Irritation, Its Principles and Practice : Illustrated by One Hundred Cases of the Most Painful and Important Diseases Effectually Cured by External Applications*. London: Churchill, 1838. http://archive.org/details/counterirritatio00gran.

Graves. "Clinical Lectures on Medicine - Acupuncture in Anasarca and Ascites." *London Medical Gazette*, NS, 1 (1839): 103–5.

Haime, A. "Notice Sur l'Acupuncture et Observations Médicales Sur Ses Effets Thérapeutiques." *Journal Universal Des Sciences Médicales* 13 (1819): 27–42.

Harris, T. "Cases of Neuralgia Treated by Galvanism." *American Journal of the Medical Sciences*, August 1834, 384.

Hippocrates. *Hippocrates - 5. Affections, Diseases I & II*. Vol. 5. Loeb Classical Library L472. Cambridge, MA: HUP, 1988. http://archive.org/details/L472HippocratesV.Affections DiseasesIII.

Kaempfer, Engelbert. *Amoenitates Exoticae, Politico-Physico - Medicae*. Amoenitates Exoticae, Politico-Physico - Medicae. Lemgoviae, 1712. https://books.google.com/books?id=Xp5Zy0O01l0C.

———. "De la Cure de la Colique par la piquure d'une Aiguille, telle qu'elle est en usage parmis les Japonnais." In *Histoire naturelle, civile et ecclésiastique du Japon,*

translated by J. G. Scheuchzer, 3:274–82. Amsterdam: Herman Uttwere, 1732.
http://archive.org/details/histoirenaturel16unkngoog.

———. "Of the Cure of the Colick by the Acupunctura or Needle-Pricking, as It Is Used by the Japanese." In *The History of Japan: Together with a Description of the Kingdom of Siam*, 3:263–72. Glasgow: MacLehose, 1906.
http://archive.org/details/historyjapantog00kaemgoog.

Kay, James Phillips. "Physiological Experiments and Observations on the Cessation of the Contractility of the Heart and Muscles in the Asphyxia of Warm-Blooded Animals." *The Edinburgh Medical and Surgical Journal* 29 (1828): 37–66.

———. *The Physiology, Pathology, and Treatment of Asphyxia*, 1834.
http://archive.org/details/physiologypatho00kaygoog.

Keate, Robert. "Treatment of Hydrocele." *London Medical Gazette*, February 1837, 789.

King. "On Acupuncturation in Ascites." *London Medical Gazette* 21 (1837): 332.

Larrey, Dominique-Jean. *On the Use of Moxa as a Therapeutical Agent. Translated from the French. With Notes and an Introduction Containing a History of the Substance, by R. Dunglison*. Translated by Robley Dunglison, 1822.
https://books.google.com/books?id=BwBeAAAAcAAJ.

Leroy. "[Asphyxied Animals]." *Journal de Physiologie* ix (n.d.).

Leroy d'Etiolles, J. "Recherches Sur L'asphyxie." *Magendie's Journal de Physiologie* vii, viii, ix (1827): 45–65.

Lewis. "New Method of Treating Hydrocele." *Lancet*, May 7, 1836.

———. "Successful Treatment of Hydrocele by Acupuncturation." *Lancet*, January 14, 1837.

Lewis, D. "New(?) Method of Treating Hydrocele; in Reply to Mr. Travers." *London Medical Gazette*, February 1837,

788.

Mansford. *Researches into the Nature and Causes of Epilepsy, &c.* Bath, 1819.

Mérat. "Percussion." In *Dictionnaire Des Sciences Médicales*, 40:288–306. Paris: Panckoucke, 1819. http://www.biusante.parisdescartes.fr/histoire/medica/resultats/?p=2&do=livre&cote=47661&fille=o&cotemere=47661.

Meyranx. "Observations Sur L'acupuncture, Faites À L'hôpital de La Pitié, Sous Les Yeux de M. Bally, et Quelques Réflexions Sur Sa Manière D'agir." *Archives Générales de Medicine* vii (1825): 231-49-96.

Most. "Galvanismus." In *Encyklopäd. Der Gesammten Medicin. Und Chirurgisch.*, 1836.

"Moxibustion." In *Dictionnaire Des Sciences Médicales*. Paris: Panckoucke, 1819. http://www.biusante.parisdescartes.fr/histoire/medica/resultats/?cote=47661x01&do=chapitre.

Pallas, Peter Simon. *Voyages de M. P. S. Pallas, en différentes provinces de l'Empire de Russie, et dans l'Asie septentrionale*. Vol. 4. Paris: Lagrange, 1788. http://gallica.bnf.fr/ark:/12148/bpt6k1040525t.

Pellatan. "Notice Sur L'acupuncture, Contenant Son Historique, Ses Effets et Sa Théorie, D'après Les Expériences Faites À L'hôpital Saint Louis [Account of the History, Effects, and Theory of Acupuncture, from Experiments Made in the Hospital of St Louis." *Revue Médicale Française et Étrangère* 1 (January 1825): 74–103.

Percy, Pierre-François. *Pyrotechnie chirurgicale pratique, ou l'art d'appliquer le feu en chirurgie*. Paris: Méquignon, 1811. https://books.google.com/books?id=10cV-lFwOD4C.

Percy, and Laurent. "Moxibustion." In *Dictionnaire Des Sciences Médicales*, n.d.

Philip, Wilson. *An Experimental Inquiry into the Laws of the Vital Functions*. London: Underwood, 1818.

http://archive.org/details/experimentalinqu01phil_0.

Pipelet. "Observation de Maladie Convulsive, Avantageusement Modifiée Par L'acupuncture." *Journal Complem. Du Dict. Des Sciences Médicales* xvi (1823): 186–87.

Pouteau, Claude. *Mélanges de chirurgie*. Paris: G. Regnault, 1760. https://books.google.com/books?id=QD9FAAAAcAAJ.

Recherches Expérimentales Sur l'Asphyxie [Galvanopuncture]. Paris, 1829.

Renton, John. "Observations on Acupuncture." *Edinburgh Medical and Surgical Journal* 34 (1830): 100–101.

Rhyne, Willem ten. *Dissertatio de arthritide: Mantissa schematica De acupunctura et orationes tres, I. De chymiae ac botaniae antiquitate & dignitate II. De physiognomia III. De monstris : singula ipsius authoris notis illustratur*. Impensis R. Chiswell, 1683. https://books.google.com/books?id=Cpi4-JIHANAC.

Riecke, V. A. *Die neuern Arzneimittel*. Hoffmann, 1837. https://books.google.com/books?id=MDJgAAAAcAAJ.

Sadler. "Indicationen Der Moxa." *Zeitschrift Für Die Gesammte Medicin.*, n.d.

———. "On the Indications for the Use of Moxa, with Cases." *British and Foreign Medical Review*, July 1837, 217.

Sarlandière, Jean Baptiste. *Mémoires sur l'électro-puncture: considérée comme moyen nouveau de traiter efficacement la goutte, les rhumatismes et les affections nerveuses, et sur l'emploi du moxa japonaia en France, suivis d'un traité de l'acupuncture et du moxa, principaux moyens curatifs chez les peuples de la Chine, de la Corée et du Japon, ornés de figurés japonaises*. Paris: Auteur et Delaunay, 1825. http://gallica.bnf.fr/ark:/12148/bpt6k5832681w/f4.image.

The Modern Part of an Universal History: From the Earliest Account of Time. Compiled from Original Writers. By the

Authors of The Antient Part. Vol. 9. London: S. Richardson, T. Osborne, C. Hitch, A. Millar, John Rivington, S. Crowder, P. Davey and B. Law, T. Longman, and C. Ware, 1759. https://books.google.com/books?id=EEwBAAAAQAAJ.

Thillaye. "Variétés." *Archives Générales de Médecine* 12 (n.d.): 461.

Travers, Benjamin. "Account of a Method of Operating for Hydrocele." *London Medical Gazette*, February 1837, 737739.

Vannes, Dantu de, and Jules Germain Cloquet. *Traité de L'acupuncture, D'après Les Observations de Jules Cloquet et Publié Sous Ses Yeux Par Dantu de Vannes [Treatise on Acupuncture, Composed from the Observations of M. Jules Cloquet, and Published under His Inspection]*. Paris: Béchet, 1826. http://www.char-fr.net/IMG/pdf/cloquet_dantu_de_vannes_traite_d_ac upuncture_1826.pdf.

Velpeau, Alfred. "Memoir on the Acupuncturation of Arteries in the Treatment of Aneurism." *London Medical Gazette* 7 (March 1830): 497–99.

———. "Mémoir Sur La Piqûre Ou L'acupuncture Des Artères Dans Le Traitement Des Anévrismes." *Gazette Médicale de Paris* 2, no. 1 (1831). https://books.google.com/books?id=PHNEAAAAcAAJ&.

———. "On the Acupuncturation of Arteries in the Treatment of Aneurism." *The American Journal of the Medical Sciences* 8 (1831): 510–12.

Velpeau, Alfred-Armand-Louis-Marie. *Traité complet d'anatomie chirurgicale, générale et topographique du corps humain, ou Anatomie considérée dans ses rapports avec la pathologie chirurgicale et la médecine opératoire*. Vol. 1. Paris: Méquignon-Marvis, 1833. http://gallica.bnf.fr/ark:/12148/bpt6k6143041d.

Vicq-d'Azyr, Félix. "Acupuncture, Acupunctura." In *Encyclopédie méthodique: ou par ordre de matières - Médecine*, 1:184–88. Paris: Panckoucke, 1787.

https://books.google.com/books?id=ZmhTAAAAcAAJ.

Von Hildenbrand. "The Anodyne Metallic or Galvanic Brush." *The Edinburgh Medical and Surgical Journal*, April 1833, 492–94.

Vowell, J.N. "Acupuncture of Ganglions." *The Lancet* 30, no. 782 (August 1838): 769–70.

Additional references

Belloc, J. J. *Cours de médecine légale judiciaire, théorique et pratique*. Paris: Méquignon, 1819. http://gallica.bnf.fr/ark:/12148/bpt6k6524171m.

Cassedy, J. "Early Uses of Acupuncture in the United States, with an Addendum (1826) by Franklin Bache, M.D." *Bull. N. Y. Acad. Med.* 50, no. 8 (1974): 892–906.

Copland, James. "[Acupuncture]." *The London Medical Repository* 3 (1825): 340–44.

Graves. "On the Treatment of Anasarca and Ascitis by Acupuncture." *The American Journal of the Medical Sciences* 23 (1839): 466–69.

Journal complémentaire des sciences médicales: recueil encyclopédique de médecine, de chirurgie et de pharmacie. Vol. 40. Paris: Panckoucke, 1831. https://books.google.com/books?id=8wQHAAAAcAAJ.

Lewis. "Acupuncture in Hydrocele and in Ascities." *London Medical Gazette* 21 (1837): 55–56.

Michel, Wolfgang. "Engelbert Kaempfers Merkwürdiger Moxa-Spiegel - Wiederholte. Lektüre Eines Deutschen Reisewerks Der Barockzeit," 1986. http://wolfgangmichel.web.fc2.com/publ/misc/index.html.

———. "The Japanese Language as Seen by Engelbert Kaempfer." *Yôgakushi Kenkyû - Western Studies* 13 (1996): 19–54.

Pouillet. "Upon the Electro-Magnetic Phenomena Which

Are Manifested in Acupuncturation." *The London Medical Repository and Review*, NS, 1, no. 3 (December 1825): 273–75.

www.ingramcontent.com/pod-product-compliance
Lightning Source LLC
Chambersburg PA
CBHW031809190326
41518CB00006B/255